P. A. Kropotkin

MUTUAL AID

A Factor of Evolution

Elibron Classics
www.elibron.com

Elibron Classics series.

ISBN 0-543-70169-7 (paperback)
ISBN 0-543-70166-2 (hardcover)

This Elibron Classics Replica Edition is an unabridged facsimile of the edition published in 1908 by William Heinemann, London.

MUTUAL AID

MUTUAL AID

A FACTOR OF EVOLUTION

BY

P. KROPOTKIN

Revised and Cheaper Edition

LONDON

WILLIAM HEINEMANN

1908

First Edition, price 7s. 6d., 1902
Second Impression, ,, ,, 1903
Revised and Cheaper Edition, 1904
Second Impression, 1907
Third Impression, March 1908

CONTENTS

INTRODUCTION

Two aspects of animal life impressed me most during the journeys which I made in my youth in Eastern Siberia and Northern Manchuria. One of them was the extreme severity of the struggle for existence which most species of animals have to carry on against an inclement Nature; the enormous destruction of life which periodically results from natural agencies; and the consequent paucity of life over the vast territory which fell under my observation. And the other was, that even in those few spots where animal life teemed in abundance, I failed to find—although I was eagerly looking for it—that bitter struggle for the means of existence, *among animals belonging to the same species*, which was considered by most Darwinists (though not always by Darwin himself) as the dominant characteristic of struggle for life, and the main factor of evolution.

The terrible snow-storms which sweep over the northern portion of Eurasia in the later part of the winter, and the glazed frost that often follows them; the frosts and the snow-storms which return every year in the second half of May, when the trees are already in full blossom and insect life swarms everywhere; the early frosts and, occasionally, the heavy snowfalls in July and August, which suddenly destroy myriads of insects, as well as the second broods of the birds in the prairies; the torrential rains, due to the monsoons, which fall in more temperate regions in August and September—resulting in inundations on a scale which is only known in America and in Eastern Asia, and swamping, on the plateaus, areas as wide as European States; and finally, the heavy snowfalls, early in October, which eventually render a territory as large as France and Germany, absolutely impracticable for ruminants, and destroy them by the thousand—these were the conditions under which I saw animal life struggling in Northern Asia. They made me realize at an early date the overwhelming importance in Nature of what Darwin described as "the natural checks to over-multiplication," in comparison to the struggle between individuals of the same species for the means of subsistence, which may go on here and there, to some limited extent, but never attains the importance of the former. Paucity of life, under-population—not over-population—being the distinctive feature of that immense part of the globe which we name Northern Asia, I conceived since then serious doubts—which subsequent study has only confirmed—as to the reality of that fearful competition for food and life within each species, which was an article of faith with most Darwinists, and, consequently, as to the dominant part which this sort of competition was supposed to play in the evolution of new species.

On the other hand, wherever I saw animal life in abundance, as, for instance, on the lakes where scores of species and millions of individuals came together to rear their progeny; in the colonies of rodents; in the migrations of birds which took place at that time on a truly American scale along the Usuri; and especially in a migration of fallow-deer which I witnessed on the Amur, and during which scores of thousands of these intelligent animals came together from an immense territory, flying before the coming deep snow, in order to cross the Amur where it is narrowest—in all these scenes of animal life which passed before my eyes, I saw Mutual Aid and Mutual Support carried on to an extent which made me suspect in it a feature of the greatest importance for the maintenance of life, the preservation of each species, and its further evolution.

And finally, I saw among the semi-wild cattle and horses in Transbaikalia, among the wild ruminants everywhere, the squirrels, and so on, that when animals have to struggle against scarcity of food, in consequence of one of the above-mentioned causes, the whole of that portion of the species which is affected by the calamity, comes out of the ordeal so much impoverished in vigour and health, that *no progressive evolution of the species can be based upon such periods of keen competition.*

Consequently, when my attention was drawn, later on, to the relations between Darwinism and Sociology, I could agree with none of the works and pamphlets that had been written upon this important subject. They all endeavoured to prove that Man, owing to his higher intelligence and knowledge, *may* mitigate the harshness of the struggle for life between men; but they all recognized at the same time that the struggle for the means of existence, of every animal against all its congeners, and of every man against all other men, was "a law of Nature." This view, however, I could not accept, because I was persuaded that to admit a pitiless inner war for life within each species, and to see in that war a condition of progress, was to admit something which not only had not yet been proved, but also lacked confirmation from direct observation.

On the contrary, a lecture "On the Law of Mutual Aid," which was delivered at a Russian Congress of Naturalists, in January 1880, by the well-known zoologist, Professor Kessler, the then Dean of the St. Petersburg University, struck me as throwing a new light on the whole subject. Kessler's idea was, that besides the *law of Mutual Struggle* there is in Nature *the law of Mutual Aid*, which, for the success of the struggle for life, and especially for the progressive evolution of the species, is far more important than the law of mutual contest. This suggestion—which was, in reality, nothing but a further development of the ideas expressed by Darwin himself in *The Descent of Man*—seemed to me so correct and of so great an importance, that since I became acquainted with it (in 1883) I began to collect materials for further developing the idea, which Kessler had only cursorily sketched in his lecture, but had not lived to develop. He died in 1881.

In one point only I could not entirely endorse Kessler's views. Kessler alluded to "parental feeling" and care for progeny (see below, Chapter I.) as to the source of mutual inclinations in animals. However, to determine how far these two feelings have really been at work in the evolution of sociable instincts, and how far other instincts have been at work in the same direction, seems to me a quite distinct and a very wide question, which we hardly can discuss yet. It will be only after we have well established the facts of mutual aid in different classes of animals, and their importance for evolution, that we shall be able to study what belongs in the evolution of sociable feelings, to parental feelings, and what to sociability proper—the latter having evidently its origin at the earliest stages of the evolution of the animal world, perhaps even at the "colony-stages." I consequently directed my chief attention to establishing first of all, the importance of the Mutual Aid factor of evolution, leaving to ulterior research the task of discovering the *origin* of the Mutual Aid instinct in Nature.

The importance of the Mutual Aid factor—"if its generality could only be demonstrated"—did not escape the naturalist's genius so manifest in Goethe. When Eckermann told once to Goethe—it was in 1827—that two little wren-fledglings, which had run away from him, were found by him next day in a nest of robin redbreasts (*Rothkehlchen*), which fed the little ones, together with their own youngsters, Goethe grew quite excited about this fact. He saw in it a confirmation of his pantheistic views, and said:—"If it be true that this feeding of a stranger goes through all Nature as something having the character of a general law—then many an enigma would be solved." He returned to this matter on the next day, and most earnestly entreated Eckermann (who was, as is known, a zoologist) to make a special study of the subject, adding that he would surely come "to quite invaluable treasuries of results" (*Gespräche*, edition of 1848, vol. iii. pp. 219, 221). Unfortunately, this study was never made, although it is very possible that Brehm, who has accumulated in his works such rich materials relative to mutual aid among animals, might have been inspired by Goethe's remark.

Several works of importance were published in the years 1872–1886, dealing with the intelligence and the mental life of animals (they are mentioned in a footnote in Chapter I. of this book), and three of them dealt more especially with the subject under consideration; namely, *Les Sociétés animales*, by Espinas (Paris, 1877); *La Lutte pour l'existence et l'association pour la lutte*, a lecture by J. L. Lanessan (April 1881); and Louis Büchner's book, *Liebe und Liebes-Leben in der Thierwelt*, of which the first edition appeared in 1882 or 1883, and a second, much enlarged, in 1885. But excellent though each of these works is, they leave ample room for a work in which Mutual Aid would be considered, not only as an argument in favour of a pre-human origin of moral instincts, but also as a law of Nature and a factor of evolution. Espinas devoted his main attention to such animal societies (ants, bees) as are established upon a physiological division of labour, and though his work is full of admirable hints in all possible directions, it was written at a time when the evolution of human societies could not yet be treated with the knowledge we now possess. Lanessan's lecture has more the character of a brilliantly-laid-out general plan of a work, in which mutual support would be dealt with, beginning with rocks in the sea, and then passing in review the world of plants, of animals and men. As to Büchner's work, suggestive though it is and rich in facts, I could not agree with its leading idea. The book begins with a hymn to Love, and nearly all its illustrations are intended to prove the existence of love and sympathy among animals. However, to reduce animal sociability to *love* and *sympathy* means to reduce its generality and its importance, just as human ethics based upon love and personal sympathy only have contributed to narrow the comprehension of the moral feeling as a whole.

It is not love to my neighbour—whom I often do not know at all—which induces me to seize a pail of water and to rush towards his house when I see it on fire; it is a far wider, even though more vague feeling or instinct of human solidarity and sociability which moves me. So it is also with animals. It is not love, and not even sympathy (understood in its proper sense) which induces a herd of ruminants or of horses to form a ring in order to resist an attack of wolves; not love which induces wolves to form a pack for hunting; not love which induces kittens or lambs to play, or a dozen of species of young birds to spend their days together in the autumn; and it is neither love nor personal sympathy which induces many thousand fallow-deer scattered over a territory as large as France to form into a score of separate herds, all marching towards a given spot, in order to cross there a river. It is a feeling infinitely wider than love or personal sympathy—an instinct that has been slowly developed among animals and men in the course of an extremely long evolution, and which has taught animals and men alike the force they can borrow from the practice of mutual aid and support, and the joys they can find in social life.

The importance of this distinction will be easily appreciated by the student of animal psychology, and the more so by the student of human ethics. Love, sympathy and self-sacrifice certainly play an immense part in the progressive development of our moral feelings. But it is not love and not even sympathy upon which Society is based in mankind. It is the conscience—be it only at the stage of an instinct—of human solidarity. It is the unconscious recognition of the force that is borrowed by each man from the practice of mutual aid; of the close dependency of every one's happiness upon the happiness of all; and of the sense of justice, or equity, which brings the individual to consider the rights of every other individual as equal to his own. Upon this broad and necessary foundation the still higher moral feelings are developed. But this subject lies outside the scope of the present work, and I shall only indicate here a lecture, "Justice and Morality," which I delivered in reply to Huxley's *Ethics*, and in which the subject has been treated at some length.

Consequently I thought that a book, written on *Mutual Aid as a Law of Nature and a factor of evolution*, might fill an important gap. When Huxley issued, in 1888, his "Struggle-for-life" manifesto (*Struggle for Existence and its Bearing upon Man*), which to my appreciation was a very incorrect representation of the facts of Nature, as one sees them in the bush and in the forest, I communicated with the editor of the *Nineteenth Century*, asking him whether he would give the hospitality of his review to an elaborate reply to the views of one of the most prominent Darwinists; and Mr. James Knowles received the proposal with fullest sympathy. I also spoke of it to W. Bates. "Yes, certainly; *that* is true Darwinism," was his reply. "It is horrible what 'they' have made of Darwin. Write these articles, and when they are printed, I will write to you a letter which you may publish." Unfortunately, it took me nearly seven years to write these articles, and when the last was published, Bates was no longer living.

After having discussed the importance of mutual aid in various classes of animals, I was evidently bound to discuss the importance of the same factor in the evolution of Man. This was the more necessary as there are a number of evolutionists who may not refuse to admit the importance of mutual aid among animals, but who, like Herbert Spencer, will refuse to admit it for Man. For primitive Man—they maintain—war of each against all was *the* law of life. In how far this assertion, which has been too willingly repeated, without sufficient criticism, since the times of Hobbes, is supported by what we know about the early phases of human development, is discussed in the chapters given to the Savages and the Barbarians.

The number and importance of mutual-aid institutions which were developed by the creative genius of the savage and half-savage masses, during the earliest clan-period of mankind and still more during the next village-community period, and the immense influence which these early institutions have exercised upon the subsequent development of mankind, down to the present times, induced me to extend my researches to the later, historical periods as well; especially, to study that most interesting period—the free mediæval city-republics, of which the universality and influence upon our modern civilization have not yet been duly appreciated. And finally, I have tried to indicate in brief the immense importance which the mutual-support instincts, inherited by mankind from its extremely long evolution, play even now in our modern society, which is supposed to rest upon the principle: "every one for himself, and the State for all," but which it never has succeeded, nor will succeed in realizing.

It may be objected to this book that both animals and men are represented in it under too favourable an aspect; that their sociable qualities are insisted upon, while their anti-social and self-asserting instincts are hardly touched upon. This was, however, unavoidable. We have heard so much lately of the "harsh, pitiless struggle for life," which was said to be carried on by every animal against all other animals, every "savage" against all other "savages," and every civilized man against all his co-citizens—and these assertions have so much become an article of faith—that it was necessary, first of all, to oppose to them a wide series of facts showing animal and human life under a quite different aspect. It was necessary to indicate the overwhelming importance which sociable habits play in Nature and in the progressive evolution of both the animal species and human beings: to prove that they secure to animals a better protection from their enemies, very often facilities for getting food (winter provisions, migrations, etc.), longevity, and therefore a greater facility for the development of intellectual faculties; and that they have given to men, in addition to the same advantages, the possibility of working out those institutions which have enabled mankind to survive in its hard struggle against Nature, and to progress, notwithstanding all the vicissitudes of its history. It is a book on the law of Mutual Aid, viewed at as one of the chief factors of evolution—not on *all* factors of evolution and their respective values; and this first book had to be written, before the latter could become possible.

I should certainly be the last to underrate the part which the self-assertion of the individual has played in the evolution of mankind. However, this subject requires, I believe, a much deeper treatment than the one it has hitherto received. In the history of mankind, individual self-assertion has often been, and continually is, something quite different from, and far larger and deeper than, the petty, unintelligent narrow-mindedness, which, with a large class of writers, goes for "individualism" and "self-assertion." Nor have history-making individuals been limited to those whom historians have represented as heroes. My intention, consequently, is, if circumstances permit it, to discuss separately the part taken by the self-assertion of the individual in the progressive evolution of mankind. I can only make in this place the following general remark:—When the Mutual Aid institutions—the tribe, the village community, the guilds, the mediæval city—began, in the course of history, to lose their primitive character, to be invaded by parasitic growths, and thus to become hindrances to progress, the revolt of individuals against these institutions took always two different aspects. Part of those who rose up strove to purify the old institutions, or to work out a higher form of commonwealth, based upon the same Mutual Aid principles; they tried, for instance, to introduce the principle of "compensation," instead of the *lex talionis*, and later on, the pardon of offences, or a still higher ideal of equality before the human conscience, *in lieu* of "compensation," according to class-value. But at the very same time, another portion of the same individual rebels endeavoured to break down the protective institutions of mutual support, with no other intention but to increase their own wealth and their own powers. In this three-cornered contest, between the two classes of revolted individuals and the supporters of what existed, lies the real tragedy of history. But to delineate that contest, and honestly to study the part played in the evolution of mankind by each one of these three forces, would require at least as many years as it took me to write this book.

Of works dealing with nearly the same subject which have been published since the publication of my articles on Mutual Aid among Animals, I must mention *The Lowell Lectures on the Ascent of Man*, by Henry Drummond (London, 1894), and *The Origin and Growth of the Moral Instinct*, by A. Sutherland (London, 1898). Both are constructed chiefly on the lines taken in Büchner's *Love*, and in the second work the parental and familial feeling as the sole influence at work in the development of the moral feelings has been dealt with at some length. A third work dealing with man and written on similar lines is *The Principles of Sociology*, by Prof. F. A. Giddings, the first edition of which was published in 1896 at New York and London, and the leading ideas of which were sketched by the author in a pamphlet in 1894. I must leave, however, to literary critics the task of discussing the points of contact, resemblance, or divergence between these works and mine.

The different chapters of this book were published first in the *Nineteenth Century* ("Mutual Aid among Animals," in September and November 1890; "Mutual Aid among Savages," in April 1891; "Mutual Aid among the Barbarians," in January 1892; "Mutual Aid in the Mediæval City," in August and September 1894; and "Mutual Aid amongst Modern Men," in January and June 1896). In bringing them out in a book form my first intention was to embody in an Appendix the mass of materials, as well as the discussion of several secondary points, which had to be omitted in the review articles. It appeared, however, that the Appendix would double the size of the book, and I was compelled to abandon, or, at least, to postpone its publication. The present

Appendix includes the discussion of only a few points which have been the matter of scientific controversy during the last few years; and into the text I have introduced only such matter as could be introduced without altering the structure of the work.

I am glad of this opportunity for expressing to the editor of the *Nineteenth Century*, Mr. James Knowles, my very best thanks, both for the kind hospitality which he offered to these papers in his review, as soon as he knew their general idea, and the permission he kindly gave me to reprint them.

Bromley, Kent,
1902.

CHAPTER I

MUTUAL AID AMONG ANIMALS

Struggle for existence.—Mutual Aid—a law of Nature and chief factor of progressive evolution.—Invertebrates.—Ants and Bees.—Birds: Hunting and fishing associations.—Sociability.—Mutual protection among small birds.—Cranes; parrots.

THE conception of struggle for existence as a factor of evolution, introduced into science by Darwin and Wallace, has permitted us to embrace an immensely-wide range of phenomena in one single generalization, which soon became the very basis of our philosophical, biological, and sociological speculations. An immense variety of facts:—adaptations of function and structure of organic beings to their surroundings; physiological and anatomical evolution; intellectual progress, and moral development itself, which we formerly used to explain by so many different causes, were embodied by Darwin in one general conception. We understood them as continued endeavours—as a struggle against adverse circumstances—for such a development of individuals, races, species and societies, as would result in the greatest possible fulness, variety, and intensity of life. It may be that at the outset Darwin himself was not fully aware of the generality of the factor which he first invoked for explaining one series only of facts relative to the accumulation of individual variations in incipient species. But he foresaw that the term which he was introducing into science would lose its philosophical and its only true meaning if it were to be used in its narrow sense only—that of a struggle between separate individuals for the sheer means of existence. And at the very beginning of his memorable work he insisted upon the term being taken in its "large and metaphorical sense including dependence of one being on another, and including (which is more important) not only the life of the individual, but success in leaving progeny."[1]

While he himself was chiefly using the term in its narrow sense for his own special purpose, he warned his followers against committing the error (which he seems once to have committed himself) of overrating its narrow meaning. In *The Descent of Man* he gave some powerful pages to illustrate its proper, wide sense. He pointed out how, in numberless animal societies, the struggle between separate individuals for the means of existence disappears, how *struggle* is replaced by *co-operation*, and how that substitution results in the development of intellectual and moral faculties which secure to the species the best conditions for survival. He intimated that in such cases the fittest are not the physically strongest, nor the cunningest, but those who learn to combine so as mutually to support each other, strong and weak alike, for the welfare of the community. "Those communities," he wrote, "which included the greatest number of the most sympathetic members would flourish best, and rear the greatest number of offspring" (2nd edit., p. 163). The term, which

[1] *Origin of Species*, chap. iii.

originated from the narrow Malthusian conception of competition between each and all, thus lost its narrowness in the mind of one who knew Nature.

Unhappily, these remarks, which might have become the basis of most fruitful researches, were overshadowed by the masses of facts gathered for the purpose of illustrating the consequences of a real competition for life. Besides, Darwin never attempted to submit to a closer investigation the relative importance of the two aspects under which the struggle for existence appears in the animal world, and he never wrote the work he proposed to write upon the natural checks to over-multiplication, although that work would have been the crucial test for appreciating the real purport of individual struggle. Nay, on the very pages just mentioned, amidst data disproving the narrow Malthusian conception of struggle, the old Malthusian leaven reappeared—namely, in Darwin's remarks as to the alleged inconveniences of maintaining the "weak in mind and body" in our civilized societies (ch. v.). As if thousands of weak-bodied and infirm poets, scientists, inventors, and reformers, together with other thousands of so-called "fools" and "weak-minded enthusiasts," were not the most precious weapons used by humanity in its struggle for existence by intellectual and moral arms, which Darwin himself emphasized in those same chapters of *Descent of Man*.

It happened with Darwin's theory as it always happens with theories having any bearing upon human relations. Instead of widening it according to his own hints, his followers narrowed it still more. And while Herbert Spencer, starting on independent but closely-allied lines, attempted to widen the inquiry into that great question, "Who are the fittest?" especially in the appendix to the third edition of the *Data of Ethics*, the numberless followers of Darwin reduced the notion of struggle for existence to its narrowest limits. They came to conceive the animal world as a world of perpetual struggle among half-starved individuals, thirsting for one another's blood. They made modern literature resound with the war-cry of *woe to the vanquished*, as if it were the last word of modern biology. They raised the "pitiless" struggle for personal advantages to the height of a biological principle which man must submit to as well, under the menace of otherwise succumbing in a world based upon mutual extermination. Leaving aside the economists who know of natural science but a few words borrowed from second-hand vulgarizers, we must recognize that even the most authorized exponents of Darwin's views did their best to maintain those false ideas. In fact, if we take Huxley, who certainly is considered as one of the ablest exponents of the theory of evolution, were we not taught by him, in a paper on the 'Struggle for Existence and its Bearing upon Man,' that,

> "from the point of view of the moralist, the animal world is on about the same level as a gladiators' show. The creatures are fairly well treated, and set to fight; whereby the strongest, the swiftest, and the cunningest live to fight another day. The spectator has no need to turn his thumb down, as no quarter is given."

Or, further down in the same article, did he not tell us that, as among animals, so among primitive men,

> "the weakest and stupidest went to the wall, while the toughest and shrewdest, those who were best fitted to cope with their circumstances, but not the best in another way, survived. Life was a continuous free fight, and beyond the limited and temporary relations of the family, the Hobbesian war of each against all was the normal state of existence." [1]

[1] *Nineteenth Century*, Feb. 1888, p. 165.

In how far this view of nature is supported by fact, will be seen from the evidence which will be here submitted to the reader as regards the animal world, and as regards primitive man. But it may be remarked at once that Huxley's view of nature had as little claim to be taken as a scientific deduction as the opposite view of Rousseau, who saw in nature but love, peace, and harmony destroyed by the accession of man. In fact, the first walk in the forest, the first observation upon any animal society, or even the perusal of any serious work dealing with animal life (D'Orbigny's, Audubon's, Le Vaillant's, no matter which), cannot but set the naturalist thinking about the part taken by social life in the life of animals, and prevent him from seeing in Nature nothing but a field of slaughter, just as this would prevent him from seeing in Nature nothing but harmony and peace. Rousseau had committed the error of excluding the beak-and-claw fight from his thoughts; and Huxley committed the opposite error; but neither Rousseau's optimism nor Huxley's pessimism can be accepted as an impartial interpretation of nature.

As soon as we study animals—not in laboratories and museums only, but in the forest and the prairie, in the steppe and the mountains—we at once perceive that though there is an immense amount of warfare and extermination going on amidst various species, and especially amidst various classes of animals, there is, at the same time, as much, or perhaps even more, of mutual support, mutual aid, and mutual defence amidst animals belonging to the same species or, at least, to the same society. Sociability is as much a law of nature as mutual struggle. Of course it would be extremely difficult to estimate, however roughly, the relative numerical importance of both these series of facts. But if we resort to an indirect test, and ask Nature: "Who are the fittest: those who are continually at war with each other, or those who support one another?" we at once see that those animals which acquire habits of mutual aid are undoubtedly the fittest. They have more chances to survive, and they attain, in their respective classes, the highest development of intelligence and bodily organization. If the numberless facts which can be brought forward to support this view are taken into account, we may safely say that mutual aid is as much a law of animal life as mutual struggle, but that, as a factor of evolution, it most probably has a far greater importance, inasmuch as it favours the development of such habits and characters as insure the maintenance and further development of the species, together with the greatest amount of welfare and enjoyment of life for the individual, with the least waste of energy.

Of the scientific followers of Darwin, the first, as far as I know, who understood the full purport of Mutual Aid *as a law of Nature and the chief factor of evolution*, was a well-known Russian zoologist, the late Dean of the St. Petersburg University, Professor Kessler. He developed his ideas in an address which he delivered in January 1880, a few months before his death, at a Congress of Russian naturalists; but, like so many good things published in the Russian tongue only, that remarkable address remains almost entirely unknown.[1]

[1] Leaving aside the pre-Darwinian writers, like Toussenel, Fée, and many others, several works containing many striking instances of mutual aid—chiefly, however, illustrating animal intelligence—were issued previously to that date. I may mention those of Houzeau, *Les facultés mentales des animaux*, 2 vols., Brussels, 1872;

"As a zoologist of old standing," he felt bound to protest against the abuse of a term—the struggle for existence—borrowed from zoology, or, at least, against overrating its importance. Zoology, he said, and those sciences which deal with man, continually insist upon what they call the pitiless law of struggle for existence. But they forget the existence of another law which may be described as the law of mutual aid, which law, at least for the animals, is far more essential than the former. He pointed out how the need of leaving progeny necessarily brings animals together, and, "the more the individuals keep together, the more they mutually support each other, and the more are the chances of the species for surviving, as well as for making further progress in its intellectual development." "All classes of animals," he continued, "and especially the higher ones, practise mutual aid," and he illustrated his idea by examples borrowed from the life of the burying beetles and the social life of birds and some mammalia. The examples were few, as

---

L. Büchner's *Aus dem Geistesleben der Thiere*, 2nd ed. in 1877; and Maximilian Perty's *Ueber das Seelenleben der Thiere*, Leipzig, 1876. Espinas published his most remarkable work, *Les Sociétés animales*, in 1877, and in that work he pointed out the importance of animal societies, and their bearing upon the preservation of species, and entered upon a most valuable discussion of the origin of societies. In fact, Espinas's book contains all that has been written since upon mutual aid, and many good things besides. If I nevertheless make a special mention of Kessler's address, it is because he raised mutual aid to the height of a law much more important in evolution than the law of mutual struggle. The same ideas were developed next year (in April 1881) by J. Lanessan in a lecture published in 1882 under this title: *La lutte pour l'existence et l'association pour la lutte.* G. Romanes's capital work, *Animal Intelligence*, was issued in 1882, and followed next year by the *Mental Evolution in Animals.* About the same time (1883), Büchner published another work, *Liebe und Liebes-Leben in der Thierwelt*, a second edition of which was issued in 1885. The idea, as seen, was in the air.

might have been expected in a short opening address, but the chief points were clearly stated; and, after mentioning that in the evolution of mankind mutual aid played a still more prominent part, Professor Kessler concluded as follows:—

"I obviously do not deny the struggle for existence, but I maintain that the progressive development of the animal kingdom, and especially of mankind, is favoured much more by mutual support than by mutual struggle. . . . All organic beings have two essential needs: that of nutrition, and that of propagating the species. The former brings them to a struggle and to mutual extermination, while the needs of maintaining the species bring them to approach one another and to support one another. But I am inclined to think that in the evolution of the organic world—in the progressive modification of organic beings—mutual support among individuals plays a much more important part than their mutual struggle." [1]

The correctness of the above views struck most of the Russian zoologists present, and Syevertsoff, whose work is well known to ornithologists and geographers, supported them and illustrated them by a few more examples. He mentioned some of the species of falcons which have "an almost ideal organization for robbery," and nevertheless are in decay, while other species of falcons, which practise mutual help, do thrive. "Take, on the other side, a sociable bird, the duck," he said; "it is poorly organized on the whole, but it practises mutual support, and it almost invades the earth, as may be judged from its numberless varieties and species."

The readiness of the Russian zoologists to accept Kessler's views seems quite natural, because nearly all of them have had opportunities of studying the

[1] *Memoirs* (*Trudy*) *of the St. Petersburg Society of Naturalists*, vol. xi. 1880.

animal world in the wide uninhabited regions of Northern Asia and East Russia; and it is impossible to study like regions without being brought to the same ideas. I recollect myself the impression produced upon me by the animal world of Siberia when I explored the Vitim regions in the company of so accomplished a zoologist as my friend Polyakoff was. We both were under the fresh impression of the *Origin of Species*, but we vainly looked for the keen competition between animals of the same species which the reading of Darwin's work had prepared us to expect, even after taking into account the remarks of the third chapter (p. 54). We saw plenty of adaptations for struggling, very often in common, against the adverse circumstances of climate, or against various enemies, and Polyakoff wrote many a good page upon the mutual dependency of carnivores, ruminants, and rodents in their geographical distribution; we witnessed numbers of facts of mutual support, especially during the migrations of birds and ruminants; but even in the Amur and Usuri regions, where animal life swarms in abundance, facts of real competition and struggle between higher animals of the same species came very seldom under my notice, though I eagerly searched for them. The same impression appears in the works of most Russian zoologists, and it probably explains why Kessler's ideas were so welcomed by the Russian Darwinists, whilst like ideas are not in vogue amidst the followers of Darwin in Western Europe.

The first thing which strikes us as soon as we begin studying the struggle for existence under both its aspects—direct and metaphorical—is the abundance of facts of mutual aid, not only for rearing progeny, as recognized by most evolutionists, but also for the safety of the individual, and for providing it with the necessary food. With many large divisions of the animal kingdom mutual aid is the rule. Mutual aid is met with even amidst the lowest animals, and we must be prepared to learn some day, from the students of microscopical pond-life, facts of unconscious mutual support, even from the life of micro-organisms. Of course, our knowledge of the life of the invertebrates, save the termites, the ants, and the bees, is extremely limited; and yet, even as regards the lower animals, we may glean a few facts of well-ascertained co-operation. The numberless associations of locusts, vanessæ, cicindelæ, cicadæ, and so on, are practically quite unexplored; but the very fact of their existence indicates that they must be composed on about the same principles as the temporary associations of ants or bees for purposes of migration.[1] As to the beetles, we have quite well-observed facts of mutual help amidst the burying beetles (*Necrophorus*). They must have some decaying organic matter to lay their eggs in, and thus to provide their larvæ with food; but that matter must not decay very rapidly. So they are wont to bury in the ground the corpses of all kinds of small animals which they occasionally find in their rambles. As a rule, they live an isolated life, but when one of them has discovered the corpse of a mouse or of a bird, which it hardly could manage to bury itself, it calls four, six, or ten other beetles to perform the operation with united efforts; if necessary, they transport the corpse to a suitable soft ground; and they bury it in a very considerate way, without

[1] See Appendix I.

quarrelling as to which of them will enjoy the privilege of laying its eggs in the buried corpse. And when Gleditsch attached a dead bird to a cross made out of two sticks, or suspended a toad to a stick planted in the soil, the little beetles would in the same friendly way combine their intelligences to overcome the artifice of Man. The same combination of efforts has been noticed among the dung-beetles.

Even among animals standing at a somewhat lower stage of organization we may find like examples. Some land-crabs of the West Indies and North America combine in large swarms in order to travel to the sea and to deposit therein their spawn; and each such migration implies concert, co-operation, and mutual support. As to the big Molucca crab (*Limulus*), I was struck (in 1882, at the Brighton Aquarium) with the extent of mutual assistance which these clumsy animals are capable of bestowing upon a comrade in case of need. One of them had fallen upon its back in a corner of the tank, and its heavy saucepan-like carapace prevented it from returning to its natural position, the more so as there was in the corner an iron bar which rendered the task still more difficult. Its comrades came to the rescue, and for one hour's time I watched how they endeavoured to help their fellow-prisoner. They came two at once, pushed their friend from beneath, and after strenuous efforts succeeded in lifting it upright; but then the iron bar would prevent them from achieving the work of rescue, and the crab would again heavily fall upon its back. After many attempts, one of the helpers would go in the depth of the tank and bring two other crabs, which would begin with fresh forces the same pushing and lifting of their helpless comrade. We stayed in the

Aquarium for more than two hours, and, when leaving, we again came to cast a glance upon the tank: the work of rescue still continued! Since I saw that, I cannot refuse credit to the observation quoted by Dr. Erasmus Darwin—namely, that "the common crab during the moulting season stations as sentinel an unmoulted or hard-shelled individual to prevent marine enemies from injuring moulted individuals in their unprotected state."[1]

Facts illustrating mutual aid amidst the termites, the ants, and the bees are so well known to the general reader, especially through the works of Romanes, L. Büchner, and Sir John Lubbock, that I may limit my remarks to a very few hints.[2] If we take an ants' nest, we not only see that every description of work—rearing of progeny, foraging, building, rearing of aphides, and so on—is performed according to the principles of voluntary mutual aid; we must also recognize, with Forel, that the chief, the fundamental feature of the life of many species of ants is the fact and the obligation for every ant of sharing its food, already swallowed and partly digested, with every member of the community which may apply for it. Two ants belonging to two different species or to two hostile nests, when they occasionally meet together, will avoid each other. But two ants belonging to the same nest or to the same colony of nests will approach

[1] George J. Romanes's *Animal Intelligence*, 1st ed. p. 233.

[2] Pierre Huber's *Recherches sur les fourmis*, Génève, 1810; reprinted as *Les fourmis indigènes*, Génève, 1861; Forel's *Recherches sur les fourmis de la Suisse*, Zurich, 1874, and J. T. Moggridge's *Harvesting Ants and Trapdoor Spiders*, London, 1873 and 1874, ought to be in the hands of every boy and girl. See also: Blanchard's *Métamorphoses des Insectes*, Paris, 1868; J. H. Fabre's *Souvenirs entomologiques*, Paris, 1886; Ebrard's *Etudes des mœurs des fourmis*, Génève, 1864; Sir John Lubbock's *Ants, Bees, and Wasps*, and so on.

each other, exchange a few movements with the antennæ, and "if one of them is hungry or thirsty, and especially if the other has its crop full . . . it immediately asks for food." The individual thus requested never refuses; it sets apart its mandibles, takes a proper position, and regurgitates a drop of transparent fluid which is licked up by the hungry ant. Regurgitating food for other ants is so prominent a feature in the life of ants (at liberty), and it so constantly recurs both for feeding hungry comrades and for feeding larvæ, that Forel considers the digestive tube of the ants as consisting of two different parts, one of which, the posterior, is for the special use of the individual, and the other, the anterior part, is chiefly for the use of the community. If an ant which has its crop full has been selfish enough to refuse feeding a comrade, it will be treated as an enemy, or even worse. If the refusal has been made while its kinsfolk were fighting with some other species, they will fall back upon the greedy individual with greater vehemence than even upon the enemies themselves. And if an ant has not refused to feed another ant belonging to an enemy species, it will be treated by the kinsfolk of the latter as a friend. All this is confirmed by most accurate observation and decisive experiments.[1]

In that immense division of the animal kingdom which embodies more than one thousand species, and is so numerous that the Brazilians pretend that Brazil belongs to the ants, not to men, competition amidst the members of the same nest, or the colony of nests,

[1] Forel's *Recherches*, pp. 244, 275, 278. Huber's description of the process is admirable. It also contains a hint as to the possible origin of the instinct (popular edition, pp. 158, 160). See Appendix II.

does not exist. However terrible the wars between different species, and whatever the atrocities committed at war-time, mutual aid within the community, self-devotion grown into a habit, and very often self-sacrifice for the common welfare, are the rule. The ants and termites have renounced the "Hobbesian war," and they are the better for it. Their wonderful nests, their buildings, superior in relative size to those of man; their paved roads and overground vaulted galleries; their spacious halls and granaries; their corn-fields, harvesting and "malting" of grain;[1] their rational methods of nursing their eggs and larvæ, and of building special nests for rearing the aphides whom Linnæus so picturesquely described as "the cows of the ants"; and, finally, their courage, pluck, and superior intelligence—all these are the natural outcome of the mutual aid which they practise at every stage of their busy and laborious lives. That mode of life also necessarily resulted in the development of another essential feature of the life of ants: the immense development of individual initiative which, in its turn, evidently led to the development of that high and varied intelligence which cannot but strike the human observer.[2]

If we knew no other facts from animal life than what

[1] The agriculture of the ants is so wonderful that for a long time it has been doubted. The fact is now so well proved by Mr. Moggridge, Dr. Lincecum, Mr. MacCook, Col. Sykes, and Dr. Jerdon, that no doubt is possible. See an excellent summary of evidence in Mr. Romanes's work. See also *Die Pilzgaerten einiger Süd-Amerikanischen Ameisen*, by Alf. Moeller, in Schimper's *Botan. Mitth. aus den Tropen*, vi. 1893.

[2] This second principle was not recognized at once. Former observers often spoke of kings, queens, managers, and so on; but since Huber and Forel have published their minute observations, no doubt is possible as to the free scope left for every individual's initiative in whatever the ants do, including their wars.

we know about the ants and the termites, we already might safely conclude that mutual aid (which leads to mutual confidence, the first condition for courage) and individual initiative (the first condition for intellectual progress) are two factors infinitely more important than mutual struggle in the evolution of the animal kingdom. In fact, the ant thrives without having any of the "protective" features which cannot be dispensed with by animals living an isolated life. Its colour renders it conspicuous to its enemies, and the lofty nests of many species are conspicuous in the meadows and forests. It is not protected by a hard carapace, and its stinging apparatus, however dangerous when hundreds of stings are plunged into the flesh of an animal, is not of a great value for individual defence; while the eggs and larvæ of the ants are a dainty for a great number of the inhabitants of the forests. And yet the ants, in their thousands, are not much destroyed by the birds, not even by the ant-eaters, and they are dreaded by most stronger insects. When Forel emptied a bagful of ants in a meadow, he saw that "the crickets ran away, abandoning their holes to be sacked by the ants; the grasshoppers and the crickets fled in all directions; the spiders and the beetles abandoned their prey in order not to become prey themselves;" even the nests of the wasps were taken by the ants, after a battle during which many ants perished for the safety of the commonwealth. Even the swiftest insects cannot escape, and Forel often saw butterflies, gnats, flies, and so on, surprised and killed by the ants. Their force is in mutual support and mutual confidence. And if the ant—apart from the still higher developed termites—stands at the very top of the whole class of insects for its intellectual capacities; if its courage is only equalled by the most courageous vertebrates; and if its brain—to use Darwin's words—"is one of the most marvellous atoms of matter in the world, perhaps more so than the brain of man," is it not due to the fact that mutual aid has entirely taken the place of mutual struggle in the communities of ants?

The same is true as regards the bees. These small insects, which so easily might become the prey of so many birds, and whose honey has so many admirers in all classes of animals from the beetle to the bear, also have none of the protective features derived from mimicry or otherwise, without which an isolatedly-living insect hardly could escape wholesale destruction; and yet, owing to the mutual aid they practise, they obtain the wide extension which we know and the intelligence we admire. By working in common they multiply their individual forces; by resorting to a temporary division of labour combined with the capacity of each bee to perform every kind of work when required, they attain such a degree of well-being and safety as no isolated animal can ever expect to achieve however strong or well-armed it may be. In their combinations they are often more successful than man, when he neglects to take advantage of a well-planned mutual assistance. Thus, when a new swarm of bees is going to leave the hive in search of a new abode, a number of bees will make a preliminary exploration of the neighbourhood, and if they discover a convenient dwelling-place—say, an old basket, or anything of the kind—they will take possession of it, clean it, and guard it, sometimes for a whole week, till the swarm comes to settle therein. But how many human settlers will perish in new countries simply for not having understood the necessity of combining their efforts! By combining their individual intelligences they succeed in coping with adverse circumstances, even quite unforeseen and unusual, like those bees of the Paris Exhibition which fastened with their resinous propolis the shutter to a glass-plate fitted in the wall of their hive. Besides, they display none of the sanguinary proclivities and love of useless fighting with which many writers so readily endow animals. The sentries which guard the entrance to the hive pitilessly put to death the robbing bees which attempt entering the hive; but those stranger bees which come to the hive by mistake are left unmolested, especially if they come laden with pollen, or are young individuals which can easily go astray. There is no more warfare than is strictly required.

The sociability of the bees is the more instructive as predatory instincts and laziness continue to exist among the bees as well, and reappear each time that their growth is favoured by some circumstances. It is well known that there always are a number of bees which prefer a life of robbery to the laborious life of a worker; and that both periods of scarcity and periods of an unusually rich supply of food lead to an increase of the robbing class. When our crops are in and there remains but little to gather in our meadows and fields, robbing bees become of more frequent occurrence; while, on the other side, about the sugar plantations of the West Indies and the sugar refineries of Europe, robbery, laziness, and very often drunkenness become quite usual with the bees. We thus see that anti-social instincts continue to exist amidst the bees as well; but natural selection continually must eliminate them, because in the long run the practice of solidarity proves much more advantageous to the species than the development of individuals endowed with predatory inclinations. The cunningest and the shrewdest are eliminated in favour of those who understand the advantages of sociable life and mutual support.

Certainly, neither the ants, nor the bees, nor even the termites, have risen to the conception of a higher solidarity embodying the whole of the species. In that respect they evidently have not attained a degree of development which we do not find even among our political, scientific, and religious leaders. Their social instincts hardly extend beyond the limits of the hive or the nest. However, colonies of no less than two hundred nests, belonging to two different species (*Formica exsecta* and *F. pressilabris*) have been described by Forel on Mount Tendre and Mount Salève; and Forel maintains that each member of these colonies recognizes every other member of the colony, and that they all take part in common defence; while in Pennsylvania Mr. MacCook saw a whole nation of from 1,600 to 1,700 nests of the mound-making ant, all living in perfect intelligence; and Mr. Bates has described the hillocks of the termites covering large surfaces in the "campos"—some of the nests being the refuge of two or three different species, and most of them being connected by vaulted galleries or arcades.[1] Some steps towards the amalgamation of larger divisions of the species for purposes of mutual protection are thus met with even among the invertebrate animals.

Going now over to higher animals, we find far more instances of undoubtedly conscious mutual help for all possible purposes, though we must recognize at once

[1] H. W. Bates, *The Naturalist on the River Amazons*, ii. 59 *seq.*

that our knowledge even of the life of higher animals still remains very imperfect. A large number of facts have been accumulated by first-rate observers, but there are whole divisions of the animal kingdom of which we know almost nothing. Trustworthy information as regards fishes is extremely scarce, partly owing to the difficulties of observation, and partly because no proper attention has yet been paid to the subject. As to the mammalia, Kessler already remarked how little we know about their manners of life. Many of them are nocturnal in their habits; others conceal themselves underground; and those ruminants whose social life and migrations offer the greatest interest do not let man approach their herds. It is chiefly upon birds that we have the widest range of information, and yet the social life of very many species remains but imperfectly known. Still, we need not complain about the lack of well-ascertained facts, as will be seen from the following.

I need not dwell upon the associations of male and female for rearing their offspring, for providing it with food during their first steps in life, or for hunting in common; though it may be mentioned by the way that such associations are the rule even with the least sociable carnivores and rapacious birds; and that they derive a special interest from being the field upon which tenderer feelings develop even amidst otherwise most cruel animals. It may also be added that the rarity of associations larger than that of the family among the carnivores and the birds of prey, though mostly being the result of their very modes of feeding, can also be explained to some extent as a consequence of the change produced in the animal world by the rapid increase of mankind. At any rate it is worthy of note that there are species living a quite isolated life in densely-inhabited regions, while the same species, or their nearest congeners, are gregarious in uninhabited countries. Wolves, foxes, and several birds of prey may be quoted as instances in point.

However, associations which do not extend beyond the family bonds are of relatively small importance in our case, the more so as we know numbers of associations for more general purposes, such as hunting, mutual protection, and even simple enjoyment of life. Audubon already mentioned that eagles occasionally associate for hunting, and his description of the two bald eagles, male and female, hunting on the Mississippi, is well known for its graphic powers. But one of the most conclusive observations of the kind belongs to Syevertsoff. Whilst studying the fauna of the Russian Steppes, he once saw an eagle belonging to an altogether gregarious species (the white-tailed eagle, *Haliaetos albicilla*) rising high in the air; for half-an-hour it was describing its wide circles in silence when at once its piercing voice was heard. Its cry was soon answered by another eagle which approached it, and was followed by a third, a fourth, and so on, till nine or ten eagles came together and soon disappeared. In the afternoon, Syevertsoff went to the place whereto he saw the eagles flying; concealed by one of the undulations of the Steppe, he approached them, and discovered that they had gathered around the corpse of a horse. The old ones, which, as a rule, begin the meal first—such are their rules of propriety—already were sitting upon the haystacks of the neighbourhood and kept watch, while the younger ones were continuing the meal, surrounded by bands of crows. From this and like observations, Syevertsoff concluded that the white-tailed eagles combine for hunting; when they all have risen to a great height they are enabled, if they are ten, to survey an area of at least twenty-five miles square; and as soon as any one has discovered something, he warns the others.[1] Of course, it might be argued that a simple instinctive cry of the first eagle, or even its movements, would have had the same effect of bringing several eagles to the prey; but in this case there is strong evidence in favour of mutual warning, because the ten eagles came together before descending towards the prey, and Syevertsoff had later on several opportunities of ascertaining that the white-tailed eagles always assemble for devouring a corpse, and that some of them (the younger ones first) always keep watch while the others are eating. In fact, the white-tailed eagle—one of the bravest and best hunters—is a gregarious bird altogether, and Brehm says that when kept in captivity it very soon contracts an attachment to its keepers.

Sociability is a common feature with very many other birds of prey. The Brazilian kite, one of the most "impudent" robbers, is nevertheless a most sociable bird. Its hunting associations have been described by Darwin and other naturalists, and it is a fact that when it has seized upon a prey which is too big, it calls together five or six friends to carry it away. After a busy day, when these kites retire for their night-rest to a tree or to the bushes, they always gather in bands, sometimes coming together from distances of ten or more miles, and they often are joined by several other vultures, especially the percnopters, "their true friends," D'Orbigny says. In

[1] N. Syevertsoff, *Periodical Phenomena in the Life of Mammalia, Birds, and Reptiles of Voronèje*, Moscow, 1855 (in Russian).

another continent, in the Transcaspian deserts, they have, according to Zarudnyi, the same habit of nesting together. The sociable vulture, one of the strongest vultures, has received its very name from its love of society. They live in numerous bands, and decidedly enjoy society; numbers of them join in their high flights for sport. "They live in very good friendship," Le Vaillant says, "and in the same cave I sometimes found as many as three nests close together."[1] The Urubú vultures of Brazil are as, or perhaps even more, sociable than rooks.[2] The little Egyptian vultures live in close friendship. They play in bands in the air, they come together to spend the night, and in the morning they all go together to search for their food, and never does the slightest quarrel arise among them; such is the testimony of Brehm, who had plenty of opportunities of observing their life. The red-throated falcon is also met with in numerous bands in the forests of Brazil, and the kestrel (*Tinnunculus cenchris*), when it has left Europe, and has reached in the winter the prairies and forests of Asia, gathers in numerous societies. In the Steppes of South Russia it is (or rather was) so sociable that Nordmann saw them in numerous bands, with other falcons (*Falco tinnunculus*, *F. œsulon*, and *F. subbuteo*), coming together every fine afternoon about four o'clock, and enjoying their sports till late in the night. They set off flying, all at once, in a quite straight line, towards some determined point, and, having reached it, immediately returned over the same line, to repeat the same flight.[3]

[1] A. Brehm, *Life of Animals*, iii. 477; all quotations after the French edition.

[2] Bates, p. 151.

[3] *Catalogue raisonné des oiseaux de la faune pontique*, in Démidoff's

To take flights in flocks for the mere pleasure of the flight, is quite common among all sorts of birds. "In the Humber district especially," Ch. Dixon writes, "vast flights of dunlins often appear upon the mud-flats towards the end of August, and remain for the winter. . . . The movements of these birds are most interesting, as a vast flock wheels and spreads out or closes up with as much precision as drilled troops. Scattered among them are many odd stints and sanderlings and ringed-plovers." [1]

It would be quite impossible to enumerate here the various hunting associations of birds; but the fishing associations of the pelicans are certainly worthy of notice for the remarkable order and intelligence displayed by these clumsy birds. They always go fishing in numerous bands, and after having chosen an appropriate bay, they form a wide half-circle in face of the shore, and narrow it by paddling towards the shore, catching all fish that happen to be enclosed in the circle. On narrow rivers and canals they even divide into two parties, each of which draws up on a half-circle, and both paddle to meet each other, just as if two parties of men dragging two long nets should advance to capture all fish taken between the nets when both parties come to meet. As the night comes they fly to their resting-places—always the same for each flock—and no one has ever seen them fighting for the possession of either the bay or the resting-place. In South America they gather in flocks of

---

Voyage; abstracts in Brehm, iii. 360. During their migrations birds of prey often associate. One flock, which H. Seebohm saw crossing the Pyrenees, represented a curious assemblage of "eight kites, one crane, and a peregrine falcon" (*The Birds of Siberia*, 1901, p. 417).

[1] *Birds in the Northern Shires*, p. 207.

from forty to fifty thousand individuals, part of which enjoy sleep while the others keep watch, and others again go fishing.[1] And finally, I should be doing an injustice to the much-calumniated house-sparrows if I did not mention how faithfully each of them shares any food it discovers with all members of the society to which it belongs. The fact was known to the Greeks, and it has been transmitted to posterity how a Greek orator once exclaimed (I quote from memory):—"While I am speaking to you a sparrow has come to tell to other sparrows that a slave has dropped on the floor a sack of corn, and they all go there to feed upon the grain." The more, one is pleased to find this observation of old confirmed in a recent little book by Mr. Gurney, who does not doubt that the house-sparrows always inform each other as to where there is some food to steal; he says, "When a stack has been thrashed ever so far from the yard, the sparrows in the yard have always had their crops full of the grain."[2] True, the sparrows are extremely particular in keeping their domains free from the invasions of strangers; thus the sparrows of the Jardin du Luxembourg bitterly fight all other sparrows which may attempt to enjoy their turn of the garden and its visitors; but within their own communities they fully practise mutual support, though occasionally there will be of course some quarrelling even amongst the best friends.

Hunting and feeding in common is so much the habit in the feathered world that more quotations hardly would be needful: it must be considered as an

[1] Max. Perty, *Ueber das Seelenleben der Thiere* (Leipzig, 1876), pp. 87, 103.

[2] G. H. Gurney, *The House-Sparrow* (London, 1885), p. 5.

established fact. As to the force derived from such associations, it is self-evident. The strongest birds of prey are powerless in face of the associations of our smallest bird pets. Even eagles—even the powerful and terrible booted eagle, and the martial eagle, which is strong enough to carry away a hare or a young antelope in its claws—are compelled to abandon their prey to bands of those beggars the kites, which give the eagle a regular chase as soon as they see it in possession of a good prey. The kites will also give chase to the swift fishing-hawk, and rob it of the fish it has captured; but no one ever saw the kites fighting together for the possession of the prey so stolen. On the Kerguelen Island, Dr. Couës saw the *Buphagus*—the sea-hen of the sealers—pursue gulls to make them disgorge their food, while, on the other side, the gulls and the terns combined to drive away the sea-hen as soon as it came near to their abodes, especially at nesting-time.[1] The little, but extremely swift lapwings (*Vanellus cristatus*) boldly attack the birds of prey. "To see them attacking a buzzard, a kite, a crow, or an eagle, is one of the most amusing spectacles. One feels that they are sure of victory, and one sees the anger of the bird of prey. In such circumstances they perfectly support one another, and their courage grows with their numbers.[2] The lapwing has well merited the name of a "good mother" which the Greeks gave to it, for it never fails to protect other aquatic birds from the attacks of their enemies. But even the little white wagtails (*Motacilla alba*), whom we well know in our gardens and whose

[1] Dr. Elliot Couës, *Birds of the Kerguelen Island*, in Smithsonian Miscellaneous Collections, vol. xiii. No. 2, p. 11.

[2] Brehm, iv. 567.

whole length hardly attains eight inches, compel the sparrow-hawk to abandon its hunt. "I often admired their courage and agility," the old Brehm wrote, "and I am persuaded that the falcon alone is capable of capturing any of them. . . . When a band of wagtails has compelled a bird of prey to retreat, they make the air resound with their triumphant cries, and after that they separate." They thus come together for the special purpose of giving chase to their enemy, just as we see it when the whole bird-population of a forest has been raised by the news that a nocturnal bird has made its appearance during the day, and all together —birds of prey and small inoffensive singers—set to chase the stranger and make it return to its concealment.

What an immense difference between the force of a kite, a buzzard or a hawk, and such small birds as the meadow-wagtail; and yet these little birds, by their common action and courage, prove superior to the powerfully-winged and armed robbers! In Europe, the wagtails not only chase the birds of prey which might be dangerous to them, but they chase also the fishing-hawk "rather for fun than for doing it any harm;" while in India, according to Dr. Jerdon's testimony, the jackdaws chase the gowinda-kite "for simple matter of amusement." Prince Wied saw the Brazilian eagle *urubitinga* surrounded by numberless flocks of toucans and cassiques (a bird nearly akin to our rook), which mocked it. "The eagle," he adds, "usually supports these insults very quietly, but from time to time it will catch one of these mockers." In all such cases the little birds, though very much inferior in force to the bird of prey, prove superior to it by their common action.[1]

[1] As to the house-sparrows, a New Zealand observer, Mr. T. W.

However, the most striking effects of common life for the security of the individual, for its enjoyment of life, and for the development of its intellectual capacities, are seen in two great families of birds, the cranes and the parrots. The cranes are extremely sociable and live in most excellent relations, not only with their congeners, but also with most aquatic birds. Their prudence is really astonishing, so also their intelligence; they grasp the new conditions in a moment, and act accordingly. Their sentries always keep watch around a flock which is feeding or resting, and the hunters know well how difficult it is to approach them. If man has succeeded in surprising them, they will never return to the same place without having sent out one single scout first, and a party of scouts afterwards; and when the reconnoitring party returns and reports that there is no danger, a second group of scouts is sent out to verify the first report, before the whole band moves. With kindred species the cranes contract real friendship; and in captivity there is no bird, save the also sociable and highly-intelligent parrot, which enters into such real friendship with man. "It sees in man, not a master, but a friend, and endeavours to manifest it," Brehm concludes from a wide personal experience. The

---

Kirk, described as follows the attack of these "impudent" birds upon an "unfortunate" hawk:—"He heard one day a most unusual noise, as though all the small birds of the country had joined in one grand quarrel. Looking up, he saw a large hawk (*C. gouldi*—a carrion feeder) being buffeted by a flock of sparrows. They kept dashing at him in scores, and from all points at once. The unfortunate hawk was quite powerless. At last, approaching some scrub, the hawk dashed into it and remained there, while the sparrows congregated in groups round the bush, keeping up a constant chattering and noise" (Paper read before the New Zealand Institute; *Nature*, Oct. 10, 1891).

crane is in continual activity from early in the morning till late in the night; but it gives a few hours only in the morning to the task of searching its food, chiefly vegetable. All the remainder of the day is given to society life. "It picks up small pieces of wood or small stones, throws them in the air and tries to catch them; it bends its neck, opens its wings, dances, jumps, runs about, and tries to manifest by all means its good disposition of mind, and always it remains graceful and beautiful."[1] As it lives in society it has almost no enemies, and though Brehm occasionally saw one of them captured by a crocodile, he wrote that except the crocodile he knew no enemies of the crane. It eschews all of them by its proverbial prudence; and it attains, as a rule, a very old age. No wonder that for the maintenance of the species the crane need not rear a numerous offspring; it usually hatches but two eggs. As to its superior intelligence, it is sufficient to say that all observers are unanimous in recognizing that its intellectual capacities remind one very much of those of man.

The other extremely sociable bird, the parrot, stands, as known, at the very top of the whole feathered world for the development of its intelligence. Brehm has so admirably summed up the manners of life of the parrot, that I cannot do better than translate the following sentence:—

"Except in the pairing season, they live in very numerous societies or bands. They choose a place in the forest to stay there, and thence they start every morning for their hunting expeditions. The members of each band remain faithfully attached to each other, and they share in common good or

[1] Brehm, iv. 671 *seq.*

bad luck. All together they repair in the morning to a field, or to a garden, or to a tree, to feed upon fruits. They post sentries to keep watch over the safety of the whole band, and are attentive to their warnings. In case of danger, all take to flight, mutually supporting each other, and all simultaneously return to their resting-place. In a word, they always live closely united."

They enjoy society of other birds as well. In India, the jays and crows come together from many miles round, to spend the night in company with the parrots in the bamboo thickets. When the parrots start hunting, they display the most wonderful intelligence, prudence, and capacity of coping with circumstances. Take, for instance, a band of white cacadoos in Australia. Before starting to plunder a corn-field, they first send out a reconnoitring party which occupies the highest trees in the vicinity of the field, while other scouts perch upon the intermediate trees between the field and the forest and transmit the signals. If the report runs "All right," a score of cacadoos will separate from the bulk of the band, take a flight in the air, and then fly towards the trees nearest to the field. They also will scrutinize the neighbourhood for a long while, and only then will they give the signal for general advance, after which the whole band starts at once and plunders the field in no time. The Australian settlers have the greatest difficulties in beguiling the prudence of the parrots; but if man, with all his art and weapons, has succeeded in killing some of them, the cacadoos become so prudent and watchful that they henceforward baffle all stratagems.[1]

There can be no doubt that it is the practice of life in society which enables the parrots to attain that very high level of almost human intelligence and almost

[1] R. Lendenfeld, in *Der zoologische Garten*, 1889.

human feelings which we know in them. Their high intelligence has induced the best naturalists to describe some species, namely the grey parrot, as the "bird-man." As to their mutual attachment it is known that when a parrot has been killed by a hunter, the others fly over the corpse of their comrade with shrieks of complaints and "themselves fall the victims of their friendship," as Audubon said; and when two captive parrots, though belonging to two different species, have contracted mutual friendship, the accidental death of one of the two friends has sometimes been followed by the death from grief and sorrow of the other friend. It is no less evident that in their societies they find infinitely more protection than they possibly might find in any ideal development of beak and claw. Very few birds of prey or mammals dare attack any but the smaller species of parrots, and Brehm is absolutely right in saying of the parrots, as he also says of the cranes and the sociable monkeys, that they hardly have any enemies besides men; and he adds: "It is most probable that the larger parrots succumb chiefly to old age rather than die from the claws of any enemies." Only man, owing to his still more superior intelligence and weapons, also derived from association, succeeds in partially destroying them. Their very longevity would thus appear as a result of their social life. Could we not say the same as regards their wonderful memory, which also must be favoured in its development by society-life and by longevity accompanied by a full enjoyment of bodily and mental faculties till a very old age?

As seen from the above, the war of each against all is not *the* law of nature. Mutual aid is as much a law of nature as mutual struggle, and that law will become still more apparent when we have analyzed some other associations of birds and those of the mammalia. A few hints as to the importance of the law of mutual aid for the evolution of the animal kingdom have already been given in the preceding pages; but their purport will still better appear when, after having given a few more illustrations, we shall be enabled presently to draw therefrom our conclusions.

CHAPTER II

MUTUAL AID AMONG ANIMALS (*continued*)

Migrations of birds.—Breeding associations.—Autumn societies.—Mammals: small number of unsociable species.—Hunting associations of wolves, lions, etc.—Societies of rodents; of ruminants; of monkeys.—Mutual Aid in the struggle for life.—Darwin's arguments to prove the struggle for life within the species.—Natural checks to over-multiplication.—Supposed extermination of intermediate links.—Elimination of competition in Nature.

As soon as spring comes back to the temperate zone, myriads and myriads of birds which are scattered over the warmer regions of the South come together in numberless bands, and, full of vigour and joy, hasten northwards to rear their offspring. Each of our hedges, each grove, each ocean cliff, and each of the lakes and ponds with which Northern America, Northern Europe, and Northern Asia are dotted tell us at that time of the year the tale of what mutual aid means for the birds; what force, energy, and protection it confers to every living being, however feeble and defenceless it otherwise might be. Take, for instance, one of the numberless lakes of the Russian and Siberian Steppes. Its shores are peopled with myriads of aquatic birds, belonging to at least a score of different species, all living in perfect peace—all protecting one another.

"For several hundred yards from the shore the air is filled with gulls and terns, as with snow-flakes on a winter day. Thousands of plovers and sand-coursers run over the beach, searching their food, whistling, and simply enjoying life. Further on, on almost each wave, a duck is rocking, while higher up you notice the flocks of the Casarki ducks. Exuberant life swarms everywhere."[1]

And here are the robbers—the strongest, the most cunning ones, those "ideally organized for robbery." And you hear their hungry, angry, dismal cries as for hours in succession they watch the opportunity of snatching from this mass of living beings one single unprotected individual. But as soon as they approach, their presence is signalled by dozens of voluntary sentries, and hundreds of gulls and terns set to chase the robber. Maddened by hunger, the robber soon abandons his usual precautions: he suddenly dashes into the living mass; but, attacked from all sides, he again is compelled to retreat. From sheer despair he falls upon the wild ducks; but the intelligent, social birds rapidly gather in a flock and fly away if the robber is an erne; they plunge into the lake if it is a falcon; or they raise a cloud of water-dust and bewilder the assailant if it is a kite.[2] And while life continues to swarm on the lake, the robber flies away with cries of anger, and looks out for carrion, or for a young bird or a field-mouse not yet used to obey in time the warnings of its comrades. In the face of an exuberant life, the ideally-armed robber must be satisfied with the off-fall of that life.

Further north, in the Arctic archipelagoes,

"you may sail along the coast for many miles and see all the ledges, all the cliffs and corners of the mountain-sides, up to

---

[1] Syevertsoff's *Periodical Phenomena*, p. 251.
[2] Seyfferlitz, quoted by Brehm, iv. 760.

a height of from two to five hundred feet, literally covered with sea-birds, whose white breasts show against the dark rocks as if the rocks were closely sprinkled with chalk specks. The air, near and far, is, so to say, full with fowls."[1]

Each of such "bird-mountains" is a living illustration of mutual aid, as well as of the infinite variety of characters, individual and specific, resulting from social life. The oyster-catcher is renowned for its readiness to attack the birds of prey. The barge is known for its watchfulness, and it easily becomes the leader of more placid birds. The turnstone, when surrounded by comrades belonging to more energetic species, is a rather timorous bird; but it undertakes to keep watch for the security of the commonwealth when surrounded by smaller birds. Here you have the dominative swans; there, the extremely sociable kittiwake-gulls, among whom quarrels are rare and short; the prepossessing polar guillemots, which continually caress each other; the egoist she-goose, who has repudiated the orphans of a killed comrade; and, by her side, another female who adopts any one's orphans, and now paddles surrounded by fifty or sixty youngsters, whom she conducts and cares for as if they all were her own breed. Side by side with the penguins, which steal one another's eggs, you have the dotterels, whose family relations are so "charming and touching" that even passionate hunters recoil from shooting a female surrounded by her young ones; or the eider-ducks, among which (like the velvet-ducks, or the *coroyas* of the Savannahs) several females hatch together in the same nest; or the lums, which sit in turn upon a common covey. Nature is

[1] *The Arctic Voyages of A. E. Nordenskjöld*, London, 1879, p. 135. See also the powerful description of the St. Kilda Islands by Mr. Dixon (quoted by Seebohm), and nearly all books of Arctic travel.

variety itself, offering all possible varieties of characters, from the basest to the highest: and that is why she cannot be depicted by any sweeping assertion. Still less can she be judged from the moralist's point of view, because the views of the moralist are themselves a result—mostly unconscious—of the observation of Nature.[1]

Coming together at nesting-time is so common with most birds that more examples are scarcely needed. Our trees are crowned with groups of crows' nests; our hedges are full of nests of smaller birds; our farmhouses give shelter to colonies of swallows; our old towers are the refuge of hundreds of nocturnal birds; and pages might be filled with the most charming descriptions of the peace and harmony which prevail in almost all these nesting associations. As to the protection derived by the weakest birds from their unions, it is evident. That excellent observer, Dr. Couës, saw, for instance, the little cliff-swallows nesting in the immediate neighbourhood of the prairie falcon (*Falco polyargus*). The falcon had its nest on the top of one of the minarets of clay which are so common in the cañons of Colorado, while a colony of swallows nested just beneath. The little peaceful birds had no fear of their rapacious neighbour; they never let it approach to their colony. They immediately surrounded it and chased it, so that it had to make off at once.[2]

[1] See Appendix III.

[2] Elliot Couës, in *Bulletin U.S. Geol. Survey of Territories*, iv. No. 7, pp. 556, 579, etc. Among the gulls (*Larus argentatus*), Polyakoff saw on a marsh in Northern Russia, that the nesting-grounds of a very great number of these birds were always patrolled by one male, which warned the colony of the approach of danger. All birds rose in such case and attacked the enemy with great vigour.

Life in societies does not cease when the nesting period is over; it begins then in a new form. The young broods gather in societies of youngsters, generally including several species. Social life is practised at that time chiefly for its own sake—partly for security, but chiefly for the pleasures derived from it. So we see in our forests the societies formed by the young nuthatchers (*Sitta cæsia*), together with titmouses, chaffinches, wrens, tree-creepers, or some wood-peckers.[1] In Spain the swallow is met with in company with kestrels, fly-catchers, and even pigeons. In the Far West of America the young horned larks live in large societies, together with another lark (Sprague's), the skylark, the Savannah sparrow, and several species of buntings and longspurs.[2] In fact, it would be much easier to describe the species which live isolated than to simply name those species which join the autumnal societies of young birds—not for hunting or nesting purposes, but simply to enjoy life in society and to spend their time in plays and sports, after having given a few hours every day to find their daily food.

And, finally, we have that immense display of mutual aid among birds—their migrations—which I dare not even enter upon in this place. Sufficient to say that birds which have lived for months in small

---

The females, which had five or six nests together on each knoll of the marsh, kept a certain order in leaving their nests in search of food. The fledglings, which otherwise are extremely unprotected and easily become the prey of the rapacious birds, were never left alone ("Family Habits among the Aquatic Birds," in *Proceedings of the Zool. Section of St. Petersburg Soc. of Nat.*, Dec. 17, 1874).

[1] Brehm Father, quoted by A. Brehm, iv. 34 *seq.* See also White's *Natural History of Selborne*, Letter XI.

[2] Dr. Couës, *Birds of Dakota and Montana*, in *Bulletin U.S. Survey of Territories*, iv. No. 7.

bands scattered over a wide territory gather in thousands; they come together at a given place, for several days in succession, before they start, and they evidently discuss the particulars of the journey. Some species will indulge every afternoon in flights preparatory to the long passage. All wait for their tardy congeners, and finally they start in a certain well-chosen direction—a fruit of accumulated collective experience—the strongest flying at the head of the band, and relieving one another in that difficult task. They cross the seas in large bands consisting of both big and small birds, and when they return next spring they repair to the same spot, and, in most cases, each of them takes possession of the very same nest which it had built or repaired the previous year.[1]

This subject is so vast, and yet so imperfectly studied; it offers so many striking illustrations of mutual-aid habits, subsidiary to the main fact of migration—each of which would, however, require a special study—that I must refrain from entering here into more details. I can only cursorily refer to the numerous and animated gatherings of birds which take place, always on the same spot, before they begin their long journeys north or south, as also those which one sees in the north, after the birds have arrived at their breeding-places on the Yenisei or in the northern counties of England. For many days in

[1] It has often been intimated that larger birds may occasionally *transport* some of the smaller birds when they cross together the Mediterranean, but the fact still remains doubtful. On the other side, it is certain that some smaller birds join the bigger ones for migration. The fact has been noticed several times, and it was recently confirmed by L. Buxbaum at Raunheim. He saw several parties of cranes which had larks flying in the midst and on both sides of their migratory columns (*Der zoologische Garten*, 1886, p. 133).

succession—sometimes one month—they will come together every morning for one hour, before flying in search of food—perhaps discussing the spot where they are going to build their nests.[1] And if, during the migration, their columns are overtaken by a storm, birds of the most different species will be brought together by common misfortune. The birds which are not exactly migratory, but slowly move northwards and southwards with the seasons, also perform these peregrinations in flocks. So far from migrating isolately, in order to secure for each separate individual the advantages of better food or shelter which are to be found in another district—they always wait for each other, and gather in flocks, before they move north or south, in accordance with the season.[2]

Going now over to mammals, the first thing which strikes us is the overwhelming numerical predominance of social species over those few carnivores which do not associate. The plateaus, the Alpine tracts, and the Steppes of the Old and New World are stocked with herds of deer, antelopes, gazelles, fallow deer, buffaloes, wild goats and sheep, all of which are sociable animals. When the Europeans came to settle in America, they found it so densely peopled with buffaloes, that pioneers had to stop their advance when a column of migrating buffaloes came to cross the route they followed; the march past of the dense

[1] H. Seebohm and Ch. Dixon both mention this habit.

[2] The fact is well known to every field-naturalist, and with reference to England several examples may be found in Charles Dixon's *Among the Birds in Northern Shires*. The chaffinches arrive during winter in vast flocks; and about the same time, *i.e.* in November, come flocks of bramblings; redwings also frequent the same places "in similar large companies," and so on (pp. 165, 166).

column lasting sometimes for two and three days. And when the Russians took possession of Siberia they found it so densely peopled with deer, antelopes, squirrels, and other sociable animals, that the very conquest of Siberia was nothing but a hunting expedition which lasted for two hundred years; while the grass plains of Eastern Africa are still covered with herds composed of zebra, the hartebeest, and other antelopes.

Not long ago the small streams of Northern America and Northern Siberia were peopled with colonies of beavers, and up to the seventeenth century like colonies swarmed in Northern Russia. The flat lands of the four great continents are still covered with countless colonies of mice, ground-squirrels, marmots, and other rodents. In the lower latitudes of Asia and Africa the forests are still the abode of numerous families of elephants, rhinoceroses, and numberless societies of monkeys. In the far north the reindeer aggregate in numberless herds; while still further north we find the herds of the musk-oxen and numberless bands of polar foxes. The coasts of the ocean are enlivened by flocks of seals and morses; its waters, by shoals of sociable cetaceans; and even in the depths of the great plateau of Central Asia we find herds of wild horses, wild donkeys, wild camels, and wild sheep. All these mammals live in societies and nations sometimes numbering hundreds of thousands of individuals, although now, after three centuries of gunpowder civilization, we find but the *débris* of the immense aggregations of old. How trifling, in comparison with them, are the numbers of the carnivores! And how false, therefore, is the view of those who speak of the animal world as if nothing were to be seen in it but lions and hyenas plunging their bleeding teeth into the flesh of their victims! One might as well imagine that the whole of human life is nothing but a succession of war massacres.

Association and mutual aid are the rule with mammals. We find social habits even among the carnivores, and we can only name the cat tribe (lions, tigers, leopards, etc.) as a division the members of which decidedly prefer isolation to society, and are but seldom met with even in small groups. And yet, even among lions "this is a very common practice to hunt in company."[1] The two tribes of the civets (*Viverridæ*) and the weasels (*Mustelidæ*) might also be characterized by their isolated life, but it is a fact that during the last century the common weasel was more sociable than it is now; it was seen then in larger groups in Scotland and in the Unterwalden canton of Switzerland. As to the great tribe of the dogs, it is eminently sociable, and association for hunting purposes may be considered as eminently characteristic of its numerous species. It is well known, in fact, that wolves gather in packs for hunting, and Tschudi left an excellent description of how they draw up in a half-circle, surround a cow which is grazing on a mountain slope, and then, suddenly appearing with a loud barking, make it roll in the abyss.[2] Audubon, in the thirties, also saw the Labrador wolves hunting in packs, and one pack following a man to his cabin, and killing the dogs. During severe winters the packs of wolves grow so numerous as to become a danger for human settlements, as was the case in France some five-and-forty

[1] S. W. Baker, *Wild Beasts*, etc., vol. i. p. 316.
[2] Tschudi, *Thierleben der Alpenwelt*, p. 404.

years ago. In the Russian Steppes they never attack the horses otherwise than in packs; and yet they have to sustain bitter fights, during which the horses (according to Kohl's testimony) sometimes assume offensive warfare, and in such cases, if the wolves do not retreat promptly, they run the risk of being surrounded by the horses and killed by their hoofs. The prairie-wolves (*Canis latrans*) are known to associate in bands of from twenty to thirty individuals when they chase a buffalo occasionally separated from its herd.[1] Jackals, which are most courageous and may be considered as one of the most intelligent representatives of the dog tribe, always hunt in packs; thus united, they have no fear of the bigger carnivores.[2] As to the wild dogs of Asia (the *Kholzuns*, or *Dholes*), Williamson saw their large packs attacking all larger animals save elephants and rhinoceroses, and overpowering bears and tigers. Hyenas always live in societies and hunt in packs, and the hunting organizations of the painted lycaons are highly praised by Cumming. Nay, even foxes, which, as a rule, live isolated in our civilized countries, have been seen combining for hunting purposes.[3] As to the polar fox, it is—or rather was in Steller's time—one of the most sociable animals; and when one reads Steller's description of the war that was waged by Behring's unfortunate crew against these intelligent small animals, one does not know what to wonder at most: the extraordinary intelligence of the foxes and the mutual aid they displayed in digging out food concealed under cairns, or stored upon a pillar (one fox would climb on

[1] Houzeau's *Études*, ii. 463.

[2] For their hunting associations see Sir E. Tennant's *Natural History of Ceylon*, quoted in Romanes's *Animal Intelligence*, p. 432.

[3] See Emil Hüter's letter in L. Büchner's *Liebe*.

its top and throw the food to its comrades beneath), or the cruelty of man, driven to despair by the numerous packs of foxes. Even some bears live in societies where they are not disturbed by man. Thus Steller saw the black bear of Kamtchatka in numerous packs, and the polar bears are occasionally found in small groups. Even the unintelligent insectivores do not always disdain association.[1]

However, it is especially with the rodents, the ungulata, and the ruminants that we find a highly-developed practice of mutual aid. The squirrels are individualist to a great extent. Each of them builds its own comfortable nest, and accumulates its own provision. Their inclinations are towards family life, and Brehm found that a family of squirrels is never so happy as when the two broods of the same year can join together with their parents in a remote corner of a forest. And yet they maintain social relations. The inhabitants of the separate nests remain in a close intercourse, and when the pine-cones become rare in the forest they inhabit, they emigrate in bands. As to the black squirrels of the Far West, they are eminently sociable. Apart from the few hours given every day to foraging, they spend their lives in playing in numerous parties. And when they multiply too rapidly in a region, they assemble in bands, almost as numerous as those of locusts, and move southwards, devastating the forests, the fields, and the gardens; while foxes, polecats, falcons, and nocturnal birds of prey follow their thick columns and live upon the individuals remaining behind. The ground-squirrel—a closely-akin genus—is still more sociable. It is given to hoarding, and stores up in its subterranean

[1] See Appendix IV.

halls large amounts of edible roots and nuts, usually plundered by man in the autumn. According to some observers, it must know something of the joys of a miser. And yet it remains sociable. It always lives in large villages, and Audubon, who opened some dwellings of the hackee in the winter, found several individuals in the same apartment; they must have stored it with common efforts.

The large tribe of the marmots, which includes the three large genuses of *Arctomys*, *Cynomys*, and *Spermophilus*, is still more sociable and still more intelligent. They also prefer having each one its own dwelling; but they live in big villages. That terrible enemy of the crops of South Russia—the *souslik*—of which some ten millions are exterminated every year by man alone, lives in numberless colonies; and while the Russian provincial assemblies gravely discuss the means of getting rid of this enemy of society, it enjoys life in its thousands in the most joyful way. Their play is so charming that no observer could refrain from paying them a tribute of praise, and from mentioning the melodious concerts arising from the sharp whistlings of the males and the melancholic whistlings of the females, before—suddenly returning to his citizen's duties—he begins inventing the most diabolic means for the extermination of the little robbers. All kinds of rapacious birds and beasts of prey having proved powerless, the last word of science in this warfare is the inoculation of cholera! The villages of the prairie-dogs in America are one of the loveliest sights. As far as the eye can embrace the prairie, it sees heaps of earth, and on each of them a prairie-dog stands, engaged in a lively conversation with its neighbours by means of short barkings. As soon as the approach of man is signalled, all plunge in a moment into their dwellings; all have disappeared as by enchantment. But if the danger is over, the little creatures soon reappear. Whole families come out of their galleries and indulge in play. The young ones scratch one another, they worry one another, and display their gracefulness while standing upright, and in the meantime the old ones keep watch. They go visiting one another, and the beaten footpaths which connect all their heaps testify to the frequency of the visitations. In short, the best naturalists have written some of their best pages in describing the associations of the prairie-dogs of America, the marmots of the Old World, and the polar marmots of the Alpine regions. And yet, I must make, as regards the marmots, the same remark as I have made when speaking of the bees. They have maintained their fighting instincts, and these instincts reappear in captivity. But in their big associations, in the face of free Nature, the unsociable instincts have no opportunity to develop, and the general result is peace and harmony.

Even such harsh animals as the rats, which continually fight in our cellars, are sufficiently intelligent not to quarrel when they plunder our larders, but to aid one another in their plundering expeditions and migrations, and even to feed their invalids. As to the beaver-rats or musk-rats of Canada, they are extremely sociable. Audubon could not but admire "their peaceful communities, which require only being left in peace to enjoy happiness." Like all sociable animals, they are lively and playful, they easily combine with other species, and they have attained a very high degree of intellectual development. In their villages, always disposed on the shores of lakes and rivers, they take into account the changing level of water; their dome-shaped houses, which are built of beaten clay interwoven with reeds, have separate corners for organic refuse, and their halls are well carpeted at winter-time; they are warm, and, nevertheless, well ventilated. As to the beavers, which are endowed, as known, with a most sympathetic character, their astounding dams and villages, in which generations live and die without knowing of any enemies but the otter and man, so wonderfully illustrate what mutual aid can achieve for the security of the species, the development of social habits, and the evolution of intelligence, that they are familiar to all interested in animal life. Let me only remark that with the beavers, the musk-rats, and some other rodents, we already find the feature which will also be distinctive of human communities—that is, work in common.

I pass in silence the two large families which include the jerboa, the chinchilla, the *biscacha*, and the *tushkan*, or underground hare of South Russia, though all these small rodents might be taken as excellent illustrations of the pleasures derived by animals from social life.[1] Precisely, the pleasures; because it is extremely difficult to say what brings animals together—the needs of mutual protection, or simply the pleasure of feeling surrounded by their congeners. At any rate, our common hares, which do not gather in societies for life

[1] With regard to the viscacha it is very interesting to note that these highly-sociable little animals not only live peaceably together in each village, but that whole villages visit each other at nights. Sociability is thus extended to the whole species—not only to a given society, or to a nation, as we saw it with the ants. When the farmer destroys a viscacha-burrow, and buries the inhabitants under a heap of earth, other viscachas—we are told by Hudson—"come from a distance to dig out those that are buried alive" (*l.c.*, p. 311). This is a widely-known fact in La Plata, verified by the author.

in common, and which are not even endowed with intense parental feelings, cannot live without coming together for play. Dietrich de Winckell, who is considered to be among the best acquainted with the habits of hares, describes them as passionate players, becoming so intoxicated by their play that a hare has been known to take an approaching fox for a playmate.[1] As to the rabbit, it lives in societies, and its family life is entirely built upon the image of the old patriarchal family; the young ones being kept in absolute obedience to the father and even the grandfather.[2] And here we have the example of two very closely-allied species which cannot bear each other—not because they live upon nearly the same food, as like cases are too often explained, but most probably because the passionate, eminently-individualist hare cannot make friends with that placid, quiet, and submissive creature, the rabbit. Their tempers are too widely different not to be an obstacle to friendship.

Life in societies is again the rule with the large family of horses, which includes the wild horses and donkeys of Asia, the zebras, the mustangs, the *cimarrones* of the Pampas, and the half-wild horses of Mongolia and Siberia. They all live in numerous associations made up of many studs, each of which consists of a number of mares under the leadership of a male. These numberless inhabitants of the Old and the New World, badly organized on the whole for resisting both their numerous enemies and the adverse conditions of climate, would soon have disappeared

[1] *Handbuch für Jäger und Jagdberechtigte*, quoted by Brehm, ii. 223.
[2] Buffon's *Histoire Naturelle*.

from the surface of the earth were it not for their sociable spirit. When a beast of prey approaches them, several studs unite at once; they repulse the beast and sometimes chase it: and neither the wolf nor the bear, not even the lion, can capture a horse or even a zebra as long as they are not detached from the herd. When a drought is burning the grass in the prairies, they gather in herds of sometimes 10,000 individuals strong, and migrate. And when a snow-storm rages in the Steppes, each stud keeps close together, and repairs to a protected ravine. But if confidence disappears, or the group has been seized by panic, and disperses, the horses perish and the survivors are found after the storm half dying from fatigue. Union is their chief arm in the struggle for life, and man is their chief enemy. Before his increasing numbers the ancestors of our domestic horse (the *Equus Przewalskii*, so named by Polyakoff) have preferred to retire to the wildest and least accessible plateaus on the outskirts of Thibet, where they continue to live, surrounded by carnivores, under a climate as bad as that of the Arctic regions, but in a region inaccessible to man.[1]

Many striking illustrations of social life could be taken from the life of the reindeer, and especially of

[1] In connection with the horses it is worthy of notice that the quagga zebra, which never comes together with the dauw zebra, nevertheless lives on excellent terms, not only with ostriches, which are very good sentries, but also with gazelles, several species of antelopes, and gnus. We thus have a case of mutual dislike between the quagga and the dauw which cannot be explained by competition for food. The fact that the quagga lives together with ruminants feeding on the same grass as itself excludes that hypothesis, and we must look for some incompatibility of character, as in the case of the hare and the rabbit. Cf., among others, Clive Phillips-Wolley's *Big Game Shooting* (Badminton Library), which contains excellent illustrations of various species living together in East Africa.

that large division of ruminants which might include the roebucks, the fallow deer, the antelopes, the gazelles, the ibex, and, in fact, the whole of the three numerous families of the Antelopides, the Caprides, and the Ovides. Their watchfulness over the safety of their herds against attacks of carnivores; the anxiety displayed by all individuals in a herd of chamois as long as all of them have not cleared a difficult passage over rocky cliffs; the adoption of orphans; the despair of the gazelle whose mate, or even comrade of the same sex, has been killed; the plays of the youngsters, and many other features, could be mentioned. But perhaps the most striking illustration of mutual support is given by the occasional migrations of fallow deer, such as I saw once on the Amur. When I crossed the high plateau and its border ridge, the Great Khingan, on my way from Transbaikalia to Merghen, and further travelled over the high prairies on my way to the Amur, I could ascertain how thinly-peopled with fallow deer these mostly uninhabited regions are.[1] Two years later I was travelling up the Amur, and by the end of October reached the lower end of that picturesque gorge which the Amur pierces in the Dousse-alin (Little Khingan) before it enters the lowlands where it joins the Sungari. I found the Cossacks in the villages of that gorge in the greatest excitement, because thousands and thousands of fallow deer were crossing the Amur where it is narrowest, in order to reach the lowlands. For several days in succession,

[1] Our Tungus hunter, who was going to marry, and therefore was prompted by the desire of getting as many furs as he possibly could, was beating the hill-sides all day long on horseback in search of deer. His efforts were not rewarded by even so much as one fallow deer killed every day; and he was an excellent hunter.

upon a length of some forty miles up the river, the Cossacks were butchering the deer as they crossed the Amur, in which already floated a good deal of ice. Thousands were killed every day, and the exodus nevertheless continued. Like migrations were never seen either before or since, and this one must have been called for by an early and heavy snow-fall in the Great Khingan, which compelled the deer to make a desperate attempt at reaching the lowlands in the east of the Dousse mountains. Indeed, a few days later the Dousse-alin was also buried under snow two or three feet deep. Now, when one imagines the immense territory (almost as big as Great Britain) from which the scattered groups of deer must have gathered for a migration which was undertaken under the pressure of exceptional circumstances, and realizes the difficulties which had to be overcome before all the deer came to the common idea of crossing the Amur further south, where it is narrowest, one cannot but deeply admire the amount of sociability displayed by these intelligent animals. The fact is not the less striking if we remember that the buffaloes of North America displayed the same powers of combination. One saw them grazing in great numbers in the plains, but these numbers were made up by an infinity of small groups which never mixed together. And yet, when necessity arose, all groups, however scattered over an immense territory, came together and made up those immense columns, numbering hundreds of thousands of individuals, which I mentioned on a preceding page.

I also ought to say a few words at least about the "compound families" of the elephants, their mutual attachment, their deliberate ways in posting sentries, and the feelings of sympathy developed by such a life of close mutual support.[1] I might mention the sociable feelings of those disreputable creatures the wild boars, and find a word of praise for their powers of association in the case of an attack by a beast of prey.[2] The hippopotamus and the rhinoceros, too, would occupy a place in a work devoted to animal sociability. Several striking pages might be given to the sociability and mutual attachment of the seals and the walruses; and finally, one might mention the most excellent feelings existing among the sociable cetaceans. But I have to say yet a few words about the societies of monkeys, which acquire an additional interest from their being the link which will bring us to the societies of primitive men.

It is hardly needful to say that those mammals, which stand at the very top of the animal world and most approach man by their structure and intelligence, are eminently sociable. Evidently we must be prepared to meet with all varieties of character and habits in so great a division of the animal kingdom which includes hundreds of species. But, all things considered, it must be said that sociability, action in common, mutual protection, and a high development of those feelings which are the necessary outcome of social life, are characteristic of most monkeys and apes. From the smallest species to the biggest ones, sociability is a rule to which we know but a few

[1] According to Samuel W. Baker, elephants combine in larger groups than the "compound family." "I have frequently observed," he wrote, "in the portion of Ceylon known as the Park Country, the tracks of elephants in great numbers which have evidently been considerable herds that have joined together in a general retreat from a ground which they considered insecure" (*Wild Beasts and their Ways*, vol. i. p. 102).

[2] Pigs, attacked by wolves, do the same (Hudson, *l.c.*).

exceptions. The nocturnal apes prefer isolated life; the capuchins (*Cebus capucinus*), the monos, and the howling monkeys live but in small families; and the orang-outans have never been seen by A. R. Wallace otherwise than either solitary or in very small groups of three or four individuals, while the gorillas seem never to join in bands. But all the remainder of the monkey tribe—the chimpanzees, the sajous, the sakis, the mandrills, the baboons, and so on—are sociable in the highest degree. They live in great bands, and even join with other species than their own. Most of them become quite unhappy when solitary. The cries of distress of each one of the band immediately bring together the whole of the band, and they boldly repulse the attacks of most carnivores and birds of prey. Even eagles do not dare attack them. They plunder our fields always in bands—the old ones taking care for the safety of the commonwealth. The little tee-tees, whose childish sweet faces so much struck Humboldt, embrace and protect one another when it rains, rolling their tails over the necks of their shivering comrades. Several species display the greatest solicitude for their wounded, and do not abandon a wounded comrade during a retreat till they have ascertained that it is dead and that they are helpless to restore it to life. Thus James Forbes narrated in his *Oriental Memoirs* a fact of such resistance in reclaiming from his hunting party the dead body of a female monkey that one fully understands why "the witnesses of this extraordinary scene resolved never again to fire at one of the monkey race."[1] In some species several individuals will combine to overturn a stone in order to search for ants' eggs under it. The

[1] Romanes's *Animal Intelligence*, p. 472.

hamadryas not only post sentries, but have been seen making a chain for the transmission of the spoil to a safe place; and their courage is well known. Brehm's description of the regular fight which his caravan had to sustain before the hamadryas would let it resume its journey in the valley of the Mensa, in Abyssinia, has become classical.[1] The playfulness of the tailed apes and the mutual attachment which reigns in the families of chimpanzees also are familiar to the general reader. And if we find among the highest apes two species, the orang-outan and the gorilla, which are not sociable, we must remember that both—limited as they are to very small areas, the one in the heart of Africa, and the other in the two islands of Borneo and Sumatra—have all the appearance of being the last remnants of formerly much more numerous species. The gorilla at least seems to have been sociable in olden times, if the apes mentioned in the *Periplus* really were gorillas.

We thus see, even from the above brief review, that life in societies is no exception in the animal world; it is the rule, the law of Nature, and it reaches its fullest development with the higher vertebrates. Those species which live solitary, or in small families only, are relatively few, and their numbers are limited. Nay, it appears very probable that, apart from a few exceptions, those birds and mammals which are not gregarious now, were living in societies before man multiplied on the earth and waged a permanent war against them, or destroyed the sources from which

[1] Brehm, i. 82; Darwin's *Descent of Man*, ch. iii. The Kozloff expedition of 1899–1901 have also had to sustain in Northern Thibet a similar fight.

they formerly derived food. "On ne s'associe pas pour mourir," was the sound remark of Espinas; and Houzeau, who knew the animal world of some parts of America when it was not yet affected by man, wrote to the same effect.

Association is found in the animal world at all degrees of evolution; and, according to the grand idea of Herbert Spencer, so brilliantly developed in Perrier's *Colonies Animales*, colonies are at the very origin of evolution in the animal kingdom. But, in proportion as we ascend the scale of evolution, we see association growing more and more conscious. It loses its purely physical character, it ceases to be simply instinctive, it becomes reasoned. With the higher vertebrates it is periodical, or is resorted to for the satisfaction of a given want—propagation of the species, migration, hunting, or mutual defence. It even becomes occasional, when birds associate against a robber, or mammals combine, under the pressure of exceptional circumstances, to emigrate. In this last case, it becomes a voluntary deviation from habitual moods of life. The combination sometimes appears in two or more degrees—the family first, then the group, and finally the association of groups, habitually scattered, but uniting in case of need, as we saw it with the bisons and other ruminants. It also takes higher forms, guaranteeing more independence to the individual without depriving it of the benefits of social life. With most rodents the individual has its own dwelling, which it can retire to when it prefers being left alone; but the dwellings are laid out in villages and cities, so as to guarantee to all inhabitants the benefits and joys of social life. And finally, in several species, such as rats, marmots, hares, etc., sociable life is maintained notwithstanding the quarrelsome or otherwise egotistic inclinations of the isolated individual. Thus it is not imposed, as is the case with ants and bees, by the very physiological structure of the individuals; it is cultivated for the benefits of mutual aid, or for the sake of its pleasures. And this, of course, appears with all possible gradations and with the greatest variety of individual and specific characters—the very variety of aspects taken by social life being a consequence, and for us a further proof, of its generality.[1]

Sociability—that is, the need of the animal of associating with its like—the love of society for society's sake, combined with the "joy of life," only now begins to receive due attention from the zoologists.[2] We know at the present time that all animals, beginning with the ants, going on to the birds, and ending with the highest mammals, are fond of plays, wrestling, running after each other, trying to capture each other, teasing each other, and so on. And while many plays are, so to speak, a school for the proper behaviour of the young in mature life, there are others, which, apart from their utilitarian purposes, are, together with dancing and singing, mere manifestations of an excess of forces—"the joy of life," and a desire to communicate in some way or another with other individuals of

[1] The more strange was it to read in the previously-mentioned article by Huxley the following paraphrase of a well-known sentence of Rousseau: "The first men who substituted mutual peace for that of mutual war—whatever the motive which impelled them to take that step—*created society*" (*Nineteenth Century*, Feb. 1888, p. 165). Society has *not* been created by man; it is anterior to man.

[2] Such monographs as the chapter on "Music and Dancing in Nature" which we have in Hudson's *Naturalist on the La Plata*, and Carl Gross' *Play of Animals*, have already thrown a considerable light upon an instinct which is absolutely universal in Nature.

the same or of other species—in short, a manifestation of *sociability proper*, which is a distinctive feature of all the animal world.[1] Whether the feeling be fear, experienced at the appearance of a bird of prey, or "a fit of gladness" which bursts out when the animals are in good health and especially when young, or merely the desire of giving play to an excess of impressions and of vital power—the necessity of communicating impressions, of playing, of chattering, or of simply feeling the proximity of other kindred living beings pervades Nature, and is, as much as any other physiological function, a distinctive feature of life and impressionability. This need takes a higher development and attains a more beautiful expression in mammals, especially amidst their young, and still more among the birds; but it pervades all Nature, and has been fully observed by the best naturalists, including Pierre Huber, even amongst the ants, and it is evidently the same instinct which brings together the big columns of butterflies which have been referred to already.

The habit of coming together for dancing and of decorating the places where the birds habitually perform their dances is, of course, well known from the pages that Darwin gave to this subject in *The Descent of Man* (ch. xiii.). Visitors of the London Zoological Gardens also know the bower of the satin bower-bird. But this habit of dancing seems to be much more widely spread than was formerly believed, and Mr. W.

[1] Not only numerous species of birds possess the habit of assembling together—in many cases always at the same spot—to indulge in antics and dancing performances, but W. H. Hudson's experience is that nearly all mammals and birds ("probably there are really *no* exceptions") indulge frequently in more or less regular or set performances with or without sound, or composed of sound exclusively (p. 264).

Hudson gives in his master-work on La Plata the most interesting description, which must be read in the original, of complicated dances, performed by quite a number of birds: rails, jacanas, lapwings, and so on.

The habit of singing in concert, which exists in several species of birds, belongs to the same category of social instincts. It is most strikingly developed with the chakar (*Chauna chavarria*), to which the English have given the most unimaginative misnomer of "crested screamer." These birds sometimes assemble in immense flocks, and in such cases they frequently sing all in concert. W. H. Hudson found them once in countless numbers, ranged all round a pampas lake in well-defined flocks, of about 500 birds in each flock.

> "Presently," he writes, "one flock near me began singing, and continued their powerful chant for three or four minutes; when they ceased the next flock took up the strains, and after it the next, and so on, until once more the notes of the flocks on the opposite shore came floating strong and clear across the water—then passed away, growing fainter and fainter, until once more the sound approached me travelling round to my side again."

On another occasion the same writer saw a whole plain covered with an endless flock of chakars, not in close order, but scattered in pairs and small groups. About nine o'clock in the evening, "suddenly the entire multitude of birds covering the marsh for miles around burst forth in a tremendous evening song. . . . It was a concert well worth riding a hundred miles to hear."[1] It may be added that like all sociable animals, the chakar easily becomes tame and grows very attached to man. "They are mild-tempered birds, and

[1] For the choruses of monkeys, see Brehm.

very rarely quarrel"—we are told—although they are well provided with formidable weapons. Life in societies renders these weapons useless.

That life in societies is the most powerful weapon in the struggle for life, taken in its widest sense, has been illustrated by several examples on the foregoing pages, and could be illustrated by any amount of evidence, if further evidence were required. Life in societies enables the feeblest insects, the feeblest birds, and the feeblest mammals to resist, or to protect themselves from, the most terrible birds and beasts of prey; it permits longevity; it enables the species to rear its progeny with the least waste of energy and to maintain its numbers albeit a very slow birth-rate; it enables the gregarious animals to migrate in search of new abodes. Therefore, while fully admitting that force, swiftness, protective colours, cunningness, and endurance to hunger and cold, which are mentioned by Darwin and Wallace, are so many qualities making the individual, or the species, the fittest under certain circumstances, we maintain that under *any* circumstances sociability is the greatest advantage in the struggle for life. Those species which willingly or unwillingly abandon it are doomed to decay; while those animals which know best how to combine, have the greatest chances of survival and of further evolution, although they may be inferior to others in *each* of the faculties enumerated by Darwin and Wallace, save the intellectual faculty. The highest vertebrates, and especially mankind, are the best proof of this assertion. As to the intellectual faculty, while every Darwinist will agree with Darwin that it is the most powerful arm in the struggle for life, and the most powerful factor of further evolution, he also will admit that intelligence is an eminently social faculty. Language, imitation, and accumulated experience are so many elements of growing intelligence of which the unsociable animal is deprived. Therefore we find, at the top of each class of animals, the ants, the parrots, and the monkeys, all combining the greatest sociability with the highest development of intelligence. The fittest are thus the most sociable animals, and sociability appears as the chief factor of evolution, both directly, by securing the well-being of the species while diminishing the waste of energy, and indirectly, by favouring the growth of intelligence.

Moreover, it is evident that life in societies would be utterly impossible without a corresponding development of social feelings, and, especially, of a certain collective sense of justice growing to become a habit. If every individual were constantly abusing its personal advantages without the others interfering in favour of the wronged, no society-life would be possible. And feelings of justice develop, more or less, with all gregarious animals. Whatever the distance from which the swallows or the cranes come, each one returns to the nest it has built or repaired last year. If a lazy sparrow intends appropriating the nest which a comrade is building, or even steals from it a few sprays of straw, the group interferes against the lazy comrade; and it is evident that without such interference being the rule, no nesting associations of birds could exist. Separate groups of penguins have separate resting-places and separate fishing abodes, and do not fight for them. The droves of cattle in Australia have particular spots to which each group repairs to rest, and from which it never deviates; and so on.[1] We have any numbers of direct observations of the peace that prevails in the nesting associations of birds, the villages of the rodents, and the herds of grass-eaters; while, on the other side, we know of few sociable animals which so continually quarrel as the rats in our cellars do, or as the morses, which fight for the possession of a sunny place on the shore. Sociability thus puts a limit to physical struggle, and leaves room for the development of better moral feelings. The high development of parental love in all classes of animals, even with lions and tigers, is generally known. As to the young birds and mammals whom we continually see associating, sympathy —not love—attains a further development in their associations. Leaving aside the really touching facts of mutual attachment and compassion which have been recorded as regards domesticated animals and with animals kept in captivity, we have a number of well-certified facts of compassion between wild animals at liberty. Max Perty and L. Büchner have given a number of such facts.[2] J. C. Wood's narrative of a weasel which came to pick up and to carry away an injured comrade enjoys a well-merited popularity.[3] So also the observation of Captain Stansbury on his journey to Utah which is quoted by Darwin; he saw a blind pelican which was fed, and well fed, by other pelicans upon fishes which had to be brought from

[1] Haygarth, *Bush Life in Australia*, p. 58.

[2] To quote but a few instances, a wounded badger was carried away by another badger suddenly appearing on the scene; rats have been seen feeding a blind couple (*Seelenleben der Thiere*, p. 64 *seq.*). Brehm himself saw two crows feeding in a hollow tree a third crow which was wounded; its wound was several weeks old (*Hausfreund*, 1874, 715; Büchner's *Liebe*, 203). Mr. Blyth saw Indian crows feeding two or three blind comrades; and so on.

[3] *Man and Beast*, p. 344.

a distance of thirty miles.[1] And when a herd of vicunas was hotly pursued by hunters, H. A. Weddell saw more than once during his journey to Bolivia and Peru, the strong males covering the retreat of the herd and lagging behind in order to protect the retreat. As to facts of compassion with wounded comrades, they are continually mentioned by all field zoologists. Such facts are quite natural. Compassion is a necessary outcome of social life. But compassion also means a considerable advance in general intelligence and sensibility. It is the first step towards the development of higher moral sentiments. It is, in its turn, a powerful factor of further evolution.

If the views developed on the preceding pages are correct, the question necessarily arises, in how far are they consistent with the theory of struggle for life as it has been developed by Darwin, Wallace, and their followers? and I will now briefly answer this important question. First of all, no naturalist will doubt that the idea of a struggle for life carried on through organic nature is the greatest generalization of our century. Life *is* struggle; and in that struggle the fittest survive. But the answers to the questions, "By which arms is this struggle chiefly carried on?" and "Who are the fittest in the struggle?" will widely differ according to the importance given to the two different aspects of the struggle: the direct one, for food and safety among separate individuals, and the struggle which Darwin described as "metaphorical"—the struggle; very often collective, against adverse circumstances. No one will deny that

[1] L. H. Morgan, *The American Beaver*, 1868, p. 272; *Descent of Man*, ch. iv.

there is, within each species, a certain amount of real competition for food—at least, at certain periods. But the question is, whether competition is carried on to the extent admitted by Darwin, or even by Wallace; and whether this competition has played, in the evolution of the animal kingdom, the part assigned to it.

The idea which permeates Darwin's work is certainly one of real competition going on within each animal group for food, safety, and possibility of leaving an offspring. He often speaks of regions being stocked with animal life to their full capacity, and from that overstocking he infers the necessity of competition. But when we look in his work for real proofs of that competition, we must confess that we do not find them sufficiently convincing. If we refer to the paragraph entitled "Struggle for Life most severe between Individuals and Varieties of the same Species," we find in it none of that wealth of proofs and illustrations which we are accustomed to find in whatever Darwin wrote. The struggle between individuals of the same species is not illustrated under that heading by even one single instance: it is taken as granted; and the competition between closely-allied animal species is illustrated by but five examples, out of which one, at least (relating to the two species of thrushes), now proves to be doubtful.[1] But when we look for more details in order

[1] One species of swallow is said to have caused the decrease of another swallow species in North America; the recent increase of the missel-thrush in Scotland has caused the decrease of the song-thrush; the brown rat has taken the place of the black rat in Europe; in Russia the small cockroach has everywhere driven before it its greater congener; and in Australia the imported hive-bee is rapidly exterminating the small stingless bee. Two other cases, but relative to domesticated animals, are mentioned in the preceding paragraph. While recalling these same facts, A. R. Wallace remarks in a footnote relative to the Scottish thrushes: "Prof. A. Newton, however,

to ascertain how far the decrease of one species was really occasioned by the increase of the other species, Darwin, with his usual fairness, tells us :

"We can dimly see why the competition should be most severe between allied forms which fill nearly the same place in nature ; but probably in no case could we precisely say why one species has been victorious over another in the great battle of life."

As to Wallace, who quotes the same facts under a slightly-modified heading ("Struggle for Life between closely-allied Animals and Plants *often* most severe"), he makes the following remark (italics are mine), which gives quite another aspect to the facts above quoted. He says :

"In *some* cases, no doubt, there is actual war between the two, the stronger killing the weaker ; *but this is by no means necessary*, and there may be cases in which the weaker species, physically, may prevail by its power of more rapid multiplication, its better withstanding vicissitudes of climate, or its greater cunning in escaping the attacks of common enemies."

In such cases what is described as competition may be no competition at all. One species succumbs, not because it is exterminated or starved out by the other species, but because it does not well accommodate itself to new conditions, which the other does. The term "struggle for life" is again used in its metaphorical sense, and may have no other. As to the real compe-

informs me that these species do not interfere in the way here stated" (*Darwinism*, p. 34). As to the brown rat, it is known that, owing to its amphibian habits, it usually stays in the lower parts of human dwellings (low cellars, sewers, etc.), as also on the banks of canals and rivers ; it also undertakes distant migrations in numberless bands. The black rat, on the contrary, prefers staying in our dwellings themselves, under the floor, as well as in our stables and barns. It thus is much more exposed to be exterminated by man ; and we cannot maintain, with any approach to certainty, that the black rat is being either exterminated or starved out by the brown rat and not by man.

tition between individuals of the same species, which is illustrated in another place by the cattle of South America during a period of drought, its value is impaired by its being taken from among domesticated animals. Bisons emigrate in like circumstances in order to avoid competition. However severe the struggle between plants—and this is amply proved—we cannot but repeat Wallace's remark to the effect that "plants live where they can," while animals have, to a great extent, the power of choice of their abode. So that we again are asking ourselves, To what extent does competition really exist within each animal species? Upon what is the assumption based?

The same remark must be made concerning the indirect argument in favour of a severe competition and struggle for life within each species, which may be derived from the "extermination of transitional varieties," so often mentioned by Darwin. It is known that for a long time Darwin was worried by the difficulty which he saw in the absence of a long chain of intermediate forms between closely-allied species, and that he found the solution of this difficulty in the supposed extermination of the intermediate forms.[1] However, an attentive reading of the different chapters in which Darwin and Wallace speak of this subject soon brings one to the conclusion that the word "extermination" does not mean real extermination;

[1] "But it may be urged that when several closely-allied species inhabit the same territory, we surely ought to find at the present time many transitional forms. . . . By my theory these allied species are descended from a common parent; and during the process of modification, each has become adapted to the conditions of life of its own region, and has supplanted and exterminated its original parent-form and all the transitional varieties between its past and present states" (*Origin of Species*, 6th ed. p. 134); also p. 137, 296 (all paragraph "On Extinction").

the same remark which Darwin made concerning his expression: "struggle for existence," evidently applies to the word "extermination" as well. It can by no means be understood in its direct sense, but must be taken "in its metaphoric sense."

If we start from the supposition that a given area is stocked with animals to its fullest capacity, and that a keen competition for the sheer means of existence is consequently going on between all the inhabitants—each animal being compelled to fight against all its congeners in order to get its daily food—then the appearance of a new and successful variety would certainly mean in many cases (though not always) the appearance of individuals which are enabled to seize more than their fair share of the means of existence; and the result would be that those individuals would starve both the parental form which does not possess the new variation and the intermediate forms which do not possess it in the same degree. It may be that at the outset, Darwin understood the appearance of new varieties under this aspect; at least, the frequent use of the word "extermination" conveys such an impression. But both he and Wallace knew Nature too well not to perceive that this is by no means the only possible and necessary course of affairs.

If the physical and the biological conditions of a given area, the extension of the area occupied by a given species, and the habits of all the members of the latter remained unchanged—then the sudden appearance of a new variety might mean the starving out and the extermination of all the individuals which were not endowed in a sufficient degree with the new feature by which the new variety is characterized. But such a combination of conditions is precisely what we do not see in Nature. Each species is continually tending to enlarge its abode; migration to new abodes is the rule with the slow snail, as with the swift bird; physical changes are continually going on in every given area; and new varieties among animals consist in an immense number of cases—perhaps in the majority—not in the growth of new weapons for snatching the food from the mouth of its congeners—food is only one out of a hundred of various conditions of existence—but, as Wallace himself shows in a charming paragraph on the "divergence of characters" (*Darwinism*, p. 107), in forming new habits, moving to new abodes, and taking to new sorts of food. In all such cases there will be no extermination, even no competition—the new adaptation being *a relief from competition, if it ever existed;* and yet there will be, after a time, an absence of intermediate links, in consequence of a mere survival of those which are best fitted for the new conditions—as surely as under the hypothesis of extermination of the parental form. It hardly need be added that if we admit, with Spencer, all the Lamarckians, and Darwin himself, the modifying influence of the surroundings upon the species, there remains still less necessity for the extermination of the intermediate forms.

The importance of migration and of the consequent isolation of groups of animals, for the origin of new varieties and ultimately of new species, which was indicated by Moritz Wagner, was fully recognized by Darwin himself. Subsequent researches have only accentuated the importance of this factor, and they have shown how the largeness of the area occupied by a given species—which Darwin considered with full reason so important for the appearance of new varieties—can be combined with the isolation of parts of the species, in consequence of local geological changes, or of local barriers. It would be impossible to enter here into the discussion of this wide question, but a few remarks will do to illustrate the combined action of these agencies. It is known that portions of a given species will often take to a new sort of food. The squirrels, for instance, when there is a scarcity of cones in the larch forests, remove to the fir-tree forests, and this change of food has certain well-known physiological effects on the squirrels. If this change of habits does not last—if next year the cones are again plentiful in the dark larch woods—no new variety of squirrels will evidently arise from this cause. But if part of the wide area occupied by the squirrels begins to have its physical characters altered—in consequence of, let us say, a milder climate or desiccation, which both bring about an increase of the pine forests in proportion to the larch woods—and if some other conditions concur to induce the squirrels to dwell on the outskirts of the desiccating region—we shall have then a new variety, *i. e.* an incipient new species of squirrels, without there having been anything that would deserve the name of extermination among the squirrels. A larger proportion of squirrels of the new, better-adapted variety would survive every year, and the intermediate links would die *in the course of time*, without having been starved out by Malthusian competitors. This is exactly what we see going on during the great physical changes which are accomplished over large areas in Central Asia, owing to the desiccation which is going on there since the glacial period.

To take another example, it has been proved by geologists that the present wild horse (*Equus Przewalski*) has slowly been evolved during the later parts of the Tertiary and the Quaternary period, but that during this succession of ages its ancestors were *not* confined to some given, limited area of the globe. They wandered over both the Old and New World, returning, in all probability, after a time to the pastures which they had, in the course of their migrations, formerly left.[1] Consequently, if we do not find now, in Asia, all the intermediate links between the present wild horse and its Asiatic Post-Tertiary ancestors, this does not mean at all that the intermediate links have been exterminated. No such extermination has ever taken place. No exceptional mortality may even have occurred among the ancestral species: the individuals which belonged to intermediate varieties and species have died in the usual course of events—often amidst plentiful food, and their remains were buried all over the globe.

In short, if we carefully consider this matter, and carefully re-read what Darwin himself wrote upon this subject, we see that if the word "extermination" be used at all in connection with transitional varieties, it must be used in its metaphoric sense. As to "competition," this expression, too, is continually used by Darwin (see, for instance, the paragraph "On Extinction") as an image, or as a way-of-speaking, rather than with the intention of conveying the idea of a real competition between two portions of the same species for the means of existence. At any rate, the absence of intermediate forms is no argument in favour of it.

In reality, the chief argument in favour of a keen

[1] According to Madame Marie Pavloff, who has made a special study of this subject, they migrated from Asia to Africa, stayed there some time, and returned next to Asia. Whether this double migration be confirmed or not, the fact of a former extension of the ancestor of our horse over Asia, Africa, and America is settled beyond doubt.

competition for the means of existence continually going on within every animal species is—to use Professor Geddes' expression—the "arithmetical argument" borrowed from Malthus.

But this argument does not prove it at all. We might as well take a number of villages in South-East Russia, the inhabitants of which enjoy plenty of food, but have no sanitary accommodation of any kind; and seeing that for the last eighty years the birth-rate was sixty in the thousand, while the population is now what it was eighty years ago, we might conclude that there has been a terrible competition between the inhabitants. But the truth is that from year to year the population remained stationary, for the simple reason that one-third of the new-born died before reaching their sixth month of life; one-half died within the next four years, and out of each hundred born, only seventeen or so reached the age of twenty. The new-comers went away before having grown to be competitors. It is evident that if such is the case with men, it is still more the case with animals. In the feathered world the destruction of the eggs goes on on such a tremendous scale that eggs are the chief food of several species in the early summer; not to say a word of the storms, the inundations which destroy nests by the million in America, and the sudden changes of weather which are fatal to the young mammals. Each storm, each inundation, each visit of a rat to a bird's nest, each sudden change of temperature, take away those competitors which appear so terrible in theory.

As to the facts of an extremely rapid increase of horses and cattle in America, of pigs and rabbits in New Zealand, and even of wild animals imported from

Europe (where their numbers are kept down by man, not by competition), they rather seem opposed to the theory of over-population. If horses and cattle could so rapidly multiply in America, it simply proved that, however numberless the buffaloes and other ruminants were at that time in the New World, its grass-eating population was far below what the prairies could maintain. If millions of intruders have found plenty of food without starving out the former population of the prairies, we must rather conclude that the Europeans found a *want* of grass-eaters in America, not an excess. And we have good reasons to believe that want of animal population is the natural state of things all over the world, with but a few temporary exceptions to the rule. The actual numbers of animals in a given region are determined, not by the highest feeding capacity of the region, but by what it is every year under the most unfavourable conditions. So that, for that reason alone, competition hardly can be a normal condition; but other causes intervene as well to cut down the animal population below even that low standard. If we take the horses and cattle which are grazing all the winter through in the Steppes of Transbaikalia, we find them very lean and exhausted at the end of the winter. But they grow exhausted not because there is not enough food for all of them—the grass buried under a thin sheet of snow is everywhere in abundance—but because of the difficulty of getting it from beneath the snow, and this difficulty is the same for all horses alike. Besides, days of glazed frost are common in early spring, and if several such days come in succession the horses grow still more exhausted. But then comes a snow-storm, which compels the already weakened animals to remain without any food for several days, and very great numbers of them die. The losses during the spring are so severe that if the season has been more inclement than usual they are even not repaired by the new breeds—the more so as *all* horses are exhausted, and the young foals are born in a weaker condition. The numbers of horses and cattle thus always remain beneath what they otherwise might be; all the year round there is food for five or ten times as many animals, and yet their population increases extremely slowly. But as soon as the Buriate owner makes ever so small a provision of hay in the steppe, and throws it open during days of glazed frost, or heavier snow-fall, he immediately sees the increase of his herd. Almost all free grass-eating animals and many rodents in Asia and America being in very much the same conditions, we can safely say that their numbers are *not* kept down by competition; that at no time of the year they can struggle for food, and that if they never reach anything approaching to over-population, the cause is in the climate, not in competition.

The importance of natural checks to over-multiplication, and especially their bearing upon the competition hypothesis, seems never to have been taken into due account. The checks, or rather some of them, are mentioned, but their action is seldom studied in detail. However, if we compare the action of the natural checks with that of competition, we must recognize at once that the latter sustains no comparison whatever with the other checks. Thus, Mr. Bates mentions the really astounding numbers of winged ants which are destroyed during their exodus. The dead or half-dead bodies of the formica de fuego (*Myrmica sævissima*) which had been blown into the river during a gale "were heaped in a line an inch or two in height and breadth, the line continuing without interruption for miles at the edge of the water."[1] Myriads of ants are thus destroyed amidst a nature which might support a hundred times as many ants as are actually living. Dr. Altum, a German forester, who wrote a very interesting book about animals injurious to our forests, also gives many facts showing the immense importance of natural checks. He says that a succession of gales or cold and damp weather during the exodus of the pine-moth (*Bombyx pini*) destroy it to incredible amounts, and during the spring of 1871 all these moths disappeared at once, probably killed by a succession of cold nights.[2] Many like examples relative to various insects could be quoted from various parts of Europe. Dr. Altum also mentions the bird-enemies of the pine-moth, and the immense amount of its eggs destroyed by foxes; but he adds that the parasitic fungi which periodically infest it are a far more terrible enemy than any bird, because they destroy the moth over very large areas at once. As to various species of mice (*Mus sylvaticus*, *Arvicola arvalis*, and *A. agrestis*), the same author gives a long list of their enemies, but he remarks: "However, the most terrible enemies of mice are not other animals, but such sudden changes of weather as occur almost every year." Alternations of frost and warm weather destroy them in numberless quantities; "one single sudden change can reduce thousands of mice to the number of a few individuals." On the other side, a warm winter, or a winter which gradually

[1] *The Naturalist on the River Amazons*, ii. 85, 95.

[2] Dr. B. Altum, *Waldbeschädigungen durch Thiere und Gegenmittel* (Berlin, 1889), pp. 207 *seq.*

steps in, make them multiply in menacing proportions, notwithstanding every enemy; such was the case in 1876 and 1877.[1] Competition, in the case of mice, thus appears a quite trifling factor when compared with weather. Other facts to the same effect are also given as regards squirrels.

As to birds, it is well known how they suffer from sudden changes of weather. Late snow-storms are as destructive of bird-life on the English moors, as they are in Siberia; and Ch. Dixon saw the red grouse so pressed during some exceptionally severe winters, that they quitted the moors in numbers, "and we have then known them actually to be taken in the streets of Sheffield. Persistent wet," he adds, "is almost as fatal to them."

On the other side, the contagious diseases which continually visit most animal species destroy them in such numbers that the losses often cannot be repaired for many years, even with the most rapidly-multiplying animals. Thus, some sixty years ago, the *sousliks* suddenly disappeared in the neighbourhood of Sarepta, in South-Eastern Russia, in consequence of some epidemics; and for years no *sousliks* were seen in that neighbourhood. It took many years before they became as numerous as they formerly were.[2]

Like facts, all tending to reduce the importance given to competition, could be produced in numbers.[3] Of course, it might be replied, in Darwin's words, that nevertheless each organic being "at some period of its life, during some season of the year, during each generation or at intervals, has to struggle for life and

[1] Dr. B. Altum, *ut supra*, pp. 13 and 187.

[2] A. Becker in the *Bulletin de la Société des Naturalistes de Moscou*, 1889, p. 625.

[3] See Appendix V.

to suffer great destruction," and that the fittest survive during such periods of hard struggle for life. But if the evolution of the animal world were based exclusively, or even chiefly, upon the survival of the fittest during periods of calamities; if natural selection were limited in its action to periods of exceptional drought, or sudden changes of temperature, or inundations, retrogression would be the rule in the animal world. Those who survive a famine, or a severe epidemic of cholera, or small-pox, or diphtheria, such as we see them in uncivilized countries, are neither the strongest, nor the healthiest, nor the most intelligent. No progress could be based on those survivals—the less so as all survivors usually come out of the ordeal with an impaired health, like the Transbaikalian horses just mentioned, or the Arctic crews, or the garrison of a fortress which has been compelled to live for a few months on half rations, and comes out of its experience with a broken health, and subsequently shows a quite abnormal mortality. All that natural selection can do in times of calamities is to spare the individuals endowed with the greatest endurance for privations of all kinds. So it does among the Siberian horses and cattle. They *are* enduring; they can feed upon the Polar birch in case of need; they resist cold and hunger. But no Siberian horse is capable of carrying half the weight which a European horse carries with ease; no Siberian cow gives half the amount of milk given by a Jersey cow, and no natives of uncivilized countries can bear a comparison with Europeans. They may better endure hunger and cold, but their physical force is very far below that of a well-fed European, and their intellectual progress is despairingly slow. "Evil cannot be productive of good," as

Tchernyshevsky wrote in a remarkable essay upon Darwinism.[1]

Happily enough, competition is not the rule either in the animal world or in mankind. It is limited among animals to exceptional periods, and natural selection finds better fields for its activity. Better conditions are created by the *elimination of competition* by means of mutual aid and mutual support.[2] In the great struggle for life—for the greatest possible fulness and intensity of life with the least waste of energy—natural selection continually seeks out the ways precisely for avoiding competition as much as possible. The ants combine in nests and nations; they pile up their stores, they rear their cattle—and thus avoid competition; and natural selection picks out of the ants' family the species which know best how to avoid competition, with its unavoidably deleterious consequences. Most of our birds slowly move southwards as the winter comes, or gather in numberless societies and undertake long journeys—and thus avoid competition. Many rodents fall asleep when the time comes that competition should set in; while other rodents store food for the winter, and gather in large villages for obtaining the necessary protection when at work. The reindeer, when the lichens are dry in the interior of the continent, migrate towards the sea. Buffaloes cross an immense continent in order to find plenty of food. And the beavers, when they grow numerous on a

[1] *Russkaya Mysl*, Sept. 1888: "The Theory of Beneficency of Struggle for Life, being a Preface to various Treatises on Botanics, Zoology, and Human Life," by an Old Transformist.

[2] "One of the most frequent modes in which Natural Selection acts is, by adapting some individuals of a species to a somewhat different mode of life, whereby they are able to seize unappropriated places in Nature" (*Origin of Species*, p. 145)—in other words, to avoid competition.

river, divide into two parties, and go, the old ones down the river, and the young ones up the river—and avoid competition. And when animals can neither fall asleep, nor migrate, nor lay in stores, nor themselves grow their food like the ants, they do what the titmouse does, and what Wallace (*Darwinism*, ch. v.) has so charmingly described: they resort to new kinds of food—and thus, again, avoid competition.[1]

"Don't compete!—competition is always injurious to the species, and you have plenty of resources to avoid it!" That is the *tendency* of nature, not always realized in full, but always present. That is the watchword which comes to us from the bush, the forest, the river, the ocean. "Therefore combine—practise mutual aid! That is the surest means for giving to each and to all the greatest safety, the best guarantee of existence and progress, bodily, intellectual, and moral." That is what Nature teaches us; and that is what all those animals which have attained the highest position in their respective classes have done. That is also what man—the most primitive man—has been doing; and that is why man has reached the position upon which we stand now, as we shall see in the subsequent chapters devoted to mutual aid in human societies.

[1] See Appendix VI.

CHAPTER III

MUTUAL AID AMONG SAVAGES

Supposed war of each against all.—Tribal origin of human society.—Late appearance of the separate family.—Bushmen and Hottentots.—Australians, Papuas.—Eskimos, Aleoutes.—Features of savage life difficult to understand for the European.—The Dayak's conception of justice.—Common law.

THE immense part played by mutual aid and mutual support in the evolution of the animal world has been briefly analyzed in the preceding chapters. We have now to cast a glance upon the part played by the same agencies in the evolution of mankind. We saw how few are the animal species which live an isolated life, and how numberless are those which live in societies, either for mutual defence, or for hunting and storing up food, or for rearing their offspring, or simply for enjoying life in common. We also saw that, though a good deal of warfare goes on between different classes of animals, or different species, or even different tribes of the same species, peace and mutual support are the rule within the tribe or the species; and that those species which best know how to combine, and to avoid competition, have the best chances of survival and of a further progressive development. They prosper, while the unsociable species decay.

It is evident that it would be quite contrary to all that we know of nature if men were an exception to so general a rule: if a creature so defenceless as man was at his beginnings should have found his protection and his way to progress, not in mutual support, like other animals, but in a reckless competition for personal advantages, with no regard to the interests of the species. To a mind accustomed to the idea of unity in nature, such a proposition appears utterly indefensible. And yet, improbable and unphilosophical as it is, it has never found a lack of supporters. There always were writers who took a pessimistic view of mankind. They knew it, more or less superficially, through their own limited experience; they knew of history what the annalists, always watchful of wars, cruelty, and oppression, told of it, and little more besides; and they concluded that mankind is nothing but a loose aggregation of beings, always ready to fight with each other, and only prevented from so doing by the intervention of some authority.

Hobbes took that position; and while some of his eighteenth-century followers endeavoured to prove that at no epoch of its existence—not even in its most primitive condition—mankind lived in a state of perpetual warfare; that men have been sociable even in "the state of nature," and that want of knowledge, rather than the natural bad inclinations of man, brought humanity to all the horrors of its early historical life,—his idea was, on the contrary, that the so-called "state of nature" was nothing but a permanent fight between individuals, accidentally huddled together by the mere caprice of their bestial existence. True, that science has made some progress since Hobbes's time, and that we have safer ground to stand upon than the speculations of Hobbes or Rousseau. But the Hobbesian philosophy has plenty of admirers still; and we have had of late quite a school of writers who, taking possession of Darwin's terminology rather than of his leading ideas, made of it an argument in favour of Hobbes's views upon primitive man, and even succeeded in giving them a scientific appearance. Huxley, as is known, took the lead of that school, and in a paper written in 1888 he represented primitive men as a sort of tigers or lions, deprived of all ethical conceptions, fighting out the struggle for existence to its bitter end, and living a life of "continual free fight"; to quote his own words—"beyond the limited and temporary relations of the family, the Hobbesian war of each against all was the normal state of existence."[1]

It has been remarked more than once that the chief error of Hobbes, and the eighteenth-century philosophers as well, was to imagine that mankind began its life in the shape of small straggling families, something like the "limited and temporary" families of the bigger carnivores, while in reality it is now positively known that such was *not* the case. Of course, we have no direct evidence as to the modes of life of the first man-like beings. We are not yet settled even as to the time of their first appearance, geologists being inclined at present to see their traces in the pliocene, or even the miocene, deposits of the Tertiary period. But we have the indirect method which permits us to throw some light even upon that remote antiquity. A most careful investigation into the social institutions of the lowest races has been carried on during the last forty years, and it has revealed among the present institutions of primitive folk some traces of still older institutions which have long disap-

[1] *Nineteenth Century*, February 1888, p. 165.

peared, but nevertheless left unmistakable traces of their previous existence. A whole science devoted to the embryology of human institutions has thus developed in the hands of Bachofen, MacLennan, Morgan, Edward B. Tylor, Maine, Post, Kovalevsky, Lubbock, and many others. And that science has established beyond any doubt that mankind did *not* begin its life in the shape of small isolated families.

Far from being a primitive form of organization, the family is a very late product of human evolution. As far as we can go back in the palæo-ethnology of mankind, we find men living in societies—in tribes similar to those of the highest mammals; and an extremely slow and long evolution was required to bring these societies to the gentile, or clan organization, which, in its turn, had to undergo another, also very long evolution, before the first germs of family, polygamous or monogamous, could appear. Societies, bands, or tribes—not families—were thus the primitive form of organization of mankind and its earliest ancestors. That is what ethnology has come to after its painstaking researches. And in so doing it simply came to what might have been foreseen by the zoologist. None of the higher mammals, save a few carnivores and a few undoubtedly-decaying species of apes (orang-outans and gorillas), live in small families, isolatedly straggling in the woods. All others live in societies. And Darwin so well understood that isolately-living apes never could have developed into man-like beings, that he was inclined to consider man as descended from some comparatively weak *but social species*, like the chimpanzee, rather than from some stronger but unsociable species, like the gorilla.[1]

[1] *The Descent of Man*, end of ch. ii. pp. 63 and 64 of the 2nd edition.

Zoology and palæo-ethnology are thus agreed in considering that the band, not the family, was the earliest form of social life. The first human societies simply were a further development of those societies which constitute the very essence of life of the higher animals.[1]

If we now go over to positive evidence, we see that the earliest traces of man, dating from the glacial or the early post-glacial period, afford unmistakable proofs of man having lived even then in societies. Isolated finds of stone implements, even from the old stone age, are very rare; on the contrary, wherever one flint implement is discovered others are sure to be found, in most cases in very large quantities. At a time when men were dwelling in caves, or under occasionally protruding rocks, in company with mammals now extinct, and hardly succeeded in making the roughest sorts of flint hatchets, they already knew the advantages of life in societies. In the valleys of the tributaries of the Dordogne, the surface of the rocks is in some places entirely covered with caves which were inhabited by palæolithic men.[2] Sometimes the cave-dwellings are superposed in storeys,

[1] Anthropologists who fully endorse the above views as regards man nevertheless intimate, sometimes, that the apes live in polygamous families, under the leadership of "a strong and jealous male." I do not know how far that assertion is based upon conclusive observation. But the passage from Brehm's *Life of Animals*, which is sometimes referred to, can hardly be taken as very conclusive. It occurs in his general description of monkeys; but his more detailed descriptions of separate species either contradict it or do not confirm it. Even as regards the cercopithèques, Brehm is affirmative in saying that they "nearly always live in bands, and very seldom in families" (French edition, p. 59). As to other species, the very numbers of their bands, always containing many males, render the "polygamous family" more than doubtful Further observation is evidently wanted.

[2] Lubbock, *Prehistoric Times*, fifth edition, 1890.

and they certainly recall much more the nesting colonies of swallows than the dens of carnivores. As to the flint implements discovered in those caves, to use Lubbock's words, "one may say without exaggeration that they are numberless." The same is true of other palæolithic stations. It also appears from Lartet's investigations that the inhabitants of the Aurignac region in the south of France partook of tribal meals at the burial of their dead. So that men lived in societies, and had germs of a tribal worship, even at that extremely remote epoch.

The same is still better proved as regards the later part of the stone age. Traces of neolithic man have been found in numberless quantities, so that we can reconstitute his manner of life to a great extent. When the ice-cap (which must have spread from the Polar regions as far south as middle France, middle Germany, and middle Russia, and covered Canada as well as a good deal of what is now the United States) began to melt away, the surfaces freed from ice were covered, first, with swamps and marshes, and later on with numberless lakes.[1] Lakes filled all depressions of the valleys before their waters dug out those permanent channels which, during a subsequent epoch, became our rivers. And wherever we explore, in Europe, Asia, or America, the shores of the literally numberless lakes of that period, whose proper name would be the Lacustrine period, we find traces of neolithic man.

[1] That extension of the ice-cap is admitted by most of the geologists who have specially studied the glacial age. The Russian Geological Survey already has taken this view as regards Russia, and most German specialists maintain it as regards Germany. The glaciation of most of the central plateau of France will not fail to be recognized by the French geologists, when they pay more attention to the glacial deposits altogether.

They are so numerous that we can only wonder at the relative density of population at that time. The "stations" of neolithic man closely follow each other on the terraces which now mark the shores of the old lakes. And at each of those stations stone implements appear in such numbers, that no doubt is possible as to the length of time during which they were inhabited by rather numerous tribes. Whole workshops of flint implements, testifying of the numbers of workers who used to come together, have been discovered by the archæologists.

Traces of a more advanced period, already characterized by the use of some pottery, are found in the shell-heaps of Denmark. They appear, as is well known, in the shape of heaps from five to ten feet thick, from 100 to 200 feet wide, and 1,000 feet or more in length, and they are so common along some parts of the sea-coast that for a long time they were considered as natural growths. And yet they "contain *nothing* but what has been in some way or other subservient to the use of man," and they are so densely stuffed with products of human industry that, during a two days' stay at Milgaard, Lubbock dug out no less than 191 pieces of stone-implements and four fragments of pottery.[1] The very size and extension of the shell-heaps prove that for generations and generations the coasts of Denmark were inhabited by hundreds of small tribes which certainly lived as peacefully together as the Fuegian tribes, which also accumulate like shell-heaps, are living in our own times.

As to the lake-dwellings of Switzerland, which represent a still further advance in civilization, they yield still better evidence of life and work in societies.

[1] *Prehistoric Times*, pp. 232 and 242.

It is known that even during the stone age the shores of the Swiss lakes were dotted with a succession of villages, each of which consisted of several huts, and was built upon a platform supported by numberless pillars in the lake. No less than twenty-four, mostly stone age villages, were discovered along the shores of Lake Leman, thirty-two in the Lake of Constance, forty-six in the Lake of Neuchâtel, and so on; and each of them testifies to the immense amount of labour which was spent in common by the tribe, not by the family. It has even been asserted that the life of the lake-dwellers must have been remarkably free of warfare. And so it probably was, especially if we refer to the life of those primitive folk who live until the present time in similar villages built upon pillars on the sea coasts.

It is thus seen, even from the above rapid hints, that our knowledge of primitive man is not so scanty after all, and that, so far as it goes, it is rather opposed than favourable to the Hobbesian speculations. Moreover, it may be supplemented, to a great extent, by the direct observation of such primitive tribes as now stand on the same level of civilization as the inhabitants of Europe stood in prehistoric times.

That these primitive tribes which we find now are *not* degenerated specimens of mankind who formerly knew a higher civilization, as it has occasionally been maintained, has sufficiently been proved by Edward B. Tylor and Lubbock. However, to the arguments already opposed to the degeneration theory, the following may be added. Save a few tribes clustering in the less-accessible highlands, the "savages" represent a girdle which encircles the more or less civilized nations, and they occupy the extremities of our continents, most of which have retained still, or recently were bearing, an early post-glacial character. Such are the Eskimos and their congeners in Greenland, Arctic America, and Northern Siberia; and, in the Southern hemisphere, the Australians, the Papuas, the Fuegians, and, partly, the Bushmen; while within the civilized area, like primitive folk are only found in the Himalayas, the highlands of Australasia, and the plateaus of Brazil. Now it must be borne in mind that the glacial age did not come to an end at once over the whole surface of the earth. It still continues in Greenland. Therefore, at a time when the littoral regions of the Indian Ocean, the Mediterranean, or the Gulf of Mexico already enjoyed a warmer climate, and became the seats of higher civilizations, immense territories in middle Europe, Siberia, and Northern America, as well as in Patagonia, Southern Africa, and Southern Australasia, remained in early post-glacial conditions which rendered them inaccessible to the civilized nations of the torrid and sub-torrid zones. They were at that time what the terrible *urmans* of North-West Siberia are now, and their population, inaccessible to and untouched by civilization, retained the characters of early post-glacial man. Later on, when desiccation rendered these territories more suitable for agriculture, they were peopled with more civilized immigrants; and while part of their previous inhabitants were assimilated by the new settlers, another part migrated further, and settled where we find them. The territories they inhabit now are still, or recently were, sub-glacial, as to their physical features; their arts and implements are those of the neolithic age; and, notwithstanding their racial differences, and the distances which separate them, their modes of life and social institutions bear a striking likeness. So we cannot but consider them as fragments of the early post-glacial population of the now civilized area.

The first thing which strikes us as soon as we begin studying primitive folk is the complexity of the organization of marriage relations under which they are living. With most of them the family, in the sense we attribute to it, is hardly found in its germs. But they are by no means loose aggregations of men and women coming in a disorderly manner together in conformity with their momentary caprices. All of them are under a certain organization, which has been described by Morgan in its general aspects as the "gentile," or clan organization.[1]

To tell the matter as briefly as possible, there is

[1] Bachofen, *Das Mutterrecht*, Stuttgart, 1861; Lewis H. Morgan, *Ancient Society, or Researches in the Lines of Human Progress from Savagery through Barbarism to Civilization*, New York, 1877; J. F. MacLennan, *Studies in Ancient History*, 1st series, new edition, 1886; 2nd series, 1896; L. Fison and A. W. Howitt, *Kamilaroi and Kurnai*, Melbourne. These four writers—as has been very truly remarked by Giraud Teulon,—starting from different facts and different general ideas, and following different methods, have come to the same conclusion. To Bachofen we owe the notion of the maternal family and the maternal succession; to Morgan—the system of kinship, Malayan and Turanian, and a highly-gifted sketch of the main phases of human evolution; to MacLennan—the law of exogeny; and to Fison and Howitt—the cuadro, or scheme, of the conjugal societies in Australia. All four end in establishing the same fact of the tribal origin of the family. When Bachofen first drew attention to the maternal family, in his epoch-making work, and Morgan described the clan-organization,—both concurring to the almost general extension of these forms and maintaining that the marriage laws lie at the very basis of the consecutive steps of human evolution, they were accused of exaggeration. However, the most careful researches prosecuted since, by a phalanx of students of ancient law, have proved that all races of mankind bear traces of having passed through similar stages of development of marriage laws, such as we now see in force among certain savages. See the works of Post, Dargun, Kovalevsky, Lubbock, and their numerous followers: Lippert, Mucke, etc.

little doubt that mankind has passed at its beginnings through a stage which may be described as that of "communal marriage"; that is, the whole tribe had husbands and wives in common with but little regard to consanguinity. But it is also certain that some restrictions to that free intercourse were imposed at a very early period. Inter-marriage was soon prohibited between the sons of one mother and her sisters, granddaughters, and aunts. Later on it was prohibited between the sons and daughters of the same mother, and further limitations did not fail to follow. The idea of a *gens*, or clan, which embodied all presumed descendants from one stock (or rather all those who gathered in one group) was evolved, and marriage within the clan was entirely prohibited. It still remained "communal," but the wife or the husband had to be taken from another clan. And when a gens became too numerous, and subdivided into several gentes, each of them was divided into classes (usually four), and marriage was permitted only between certain well-defined classes. That is the stage which we find now among the Kamilaroi-speaking Australians. As to the family, its first germs appeared amidst the clan organization. A woman who was captured in war from some other clan, and who formerly would have belonged to the whole gens, could be kept at a later period by the capturer, under certain obligations towards the tribe. She may be taken by him to a separate hut, after she had paid a certain tribute to the clan, and thus constitute within the gens a separate family, the appearance of which evidently was opening a quite new phase of civilization.[1]

Now, if we take into consideration that this com-

[1] See Appendix VII.

plicated organization developed among men who stood at the lowest known degree of development, and that it maintained itself in societies knowing no kind of authority besides the authority of public opinion, we at once see how deeply inrooted social instincts must have been in human nature, even at its lowest stages. A savage who is capable of living under such an organization, and of freely submitting to rules which continually clash with his personal desires, certainly is not a beast devoid of ethical principles and knowing no rein to its passions. But the fact becomes still more striking if we consider the immense antiquity of the clan organization. It is now known that the primitive Semites, the Greeks of Homer, the prehistoric Romans, the Germans of Tacitus, the early Celts and the early Slavonians, all have had their own period of clan organization, closely analogous to that of the Australians, the Red Indians, the Eskimos, and other inhabitants of the "savage girdle."[1] So we must admit that either the evolution of marriage laws went on on the same lines among all human races, or the rudiments of the clan rules were developed among some common ancestors of the Semites, the Aryans, the Polynesians, etc., before their differentiation into separate races took place, and that these rules were maintained, until now, among races long ago separated from the common stock. Both alternatives imply, however, an equally striking tenacity of the

[1] For the Semites and the Aryans, see especially Prof. Maxim Kovalevsky's *Primitive Law* (in Russian), Moscow, 1886 and 1887. Also his lectures delivered at Stockholm (*Tableau des origines et de l'évolution de la famille et de la propriété*, Stockholm, 1890), which represents an admirable review of the whole question. Cf. also A. Post *Die Geschlechtsgenossenschaft der Urzeit*, Oldenburg, 1875.

institution—such a tenacity that no assaults of the individual could break it down through the scores of thousands of years that it was in existence. The very persistence of the clan organization shows how utterly false it is to represent primitive mankind as a disorderly agglomeration of individuals, who only obey their individual passions, and take advantage of their personal force and cunningness against all other representatives of the species. Unbridled individualism is a modern growth, but it is not characteristic of primitive mankind.[1]

Going now over to the existing savages, we may begin with the Bushmen, who stand at a very low level of development—so low indeed that they have no dwellings and sleep in holes dug in the soil, occasionally protected by some screens. It is known that when Europeans settled in their territory and destroyed deer, the Bushmen began stealing the settlers' cattle, whereupon a war of extermination, too horrible to be related here, was waged against them.

[1] It would be impossible to enter here into a discussion of the origin of the marriage restrictions. Let me only remark that a division into groups, similar to Morgan's *Hawaian*, exists among birds; the young broods live together separately from their parents. A like division might probably be traced among some mammals as well. As to the prohibition of relations between brothers and sisters, it is more likely to have arisen, not from speculations about the bad effects of consanguinity, which speculations really do not seem probable, but to avoid the too-easy precocity of like marriages. Under close cohabitation it must have become of imperious necessity. I must also remark that in discussing the origin of new customs altogether, we must keep in mind that the savages, like us, have their "thinkers" and *savants*—wizards, doctors, prophets, etc.—whose knowledge and ideas are in advance upon those of the masses. United as they are in their secret unions (another almost universal feature) they are certainly capable of exercising a powerful influence, and of enforcing customs the utility of which may not yet be recognized by the majority of the tribe.

Five hundred Bushmen were slaughtered in 1774, three thousand in 1808 and 1809 by the Farmers' Alliance, and so on. They were poisoned like rats, killed by hunters lying in ambush before the carcass of some animal, killed wherever met with.[1] So that our knowledge of the Bushmen, being chiefly borrowed from those same people who exterminated them, is necessarily limited. But still we know that when the Europeans came, the Bushmen lived in small tribes (or clans), sometimes federated together; that they used to hunt in common, and divided the spoil without quarrelling; that they never abandoned their wounded, and displayed strong affection to their comrades. Lichtenstein has a most touching story about a Bushman, nearly drowned in a river, who was rescued by his companions. They took off their furs to cover him, and shivered themselves; they dried him, rubbed him before the fire, and smeared his body with warm grease till they brought him back to life. And when the Bushmen found, in Johan van der Walt, a man who treated them well, they expressed their thankfulness by a most touching attachment to that man.[2] Burchell and Moffat both represent them as good-hearted, disinterested, true to their promises, and grateful,[3] all qualities which could develop only by being practised within the tribe. As to their love to children, it is sufficient to say that when a European wished to secure a Bushman woman as a slave, he

[1] Col. Collins, in Philips' *Researches in South Africa*, London, 1828. Quoted by Waitz, ii. 334.

[2] Lichtenstein's *Reisen im südlichen Afrika*, ii. pp. 92, 97. Berlin, 1811.

[3] Waitz, *Anthropologie der Naturvölker*, ii. pp. 335 *seq.* See also Fritsch's *Die Eingeboren Afrika's*, Breslau, 1872, pp. 386 *seq.*; and *Drei Jahre in Süd-Afrika.* Also W. Bleck, *A Brief Account of Bushmen Folklore*, Capetown, 1875.

stole her child: the mother was sure to come into slavery to share the fate of her child.[1]

The same social manners characterize the Hottentots, who are but a little more developed than the Bushmen. Lubbock describes them as "the filthiest animals," and filthy they really are. A fur suspended to the neck and worn till it falls to pieces is all their dress; their huts are a few sticks assembled together and covered with mats, with no kind of furniture within. And though they kept oxen and sheep, and seem to have known the use of iron before they made acquaintance with the Europeans, they still occupy one of the lowest degrees of the human scale. And yet those who knew them highly praised their sociability and readiness to aid each other. If anything is given to a Hottentot, he at once divides it among all present—a habit which, as is known, so much struck Darwin among the Fuegians. He cannot eat alone, and, however hungry, he calls those who pass by to share his food. And when Kolben expressed his astonishment thereat, he received the answer: "That is Hottentot manner." But this is not Hottentot manner only: it is an all but universal habit among the "savages." Kolben, who knew the Hottentots well and did not pass by their defects in silence, could not praise their tribal morality highly enough.

"Their word is sacred," he wrote. They know "nothing of the corruptness and faithless arts of Europe." "They live in great tranquillity and are seldom at war with their neighbours." They are "all kindness and goodwill to one another. . . . One of the greatest pleasures of the Hottentots certainly lies in their gifts and good offices to one another." "The integrity of the Hottentots, their strictness and celerity in the

---

[1] Elisée Reclus, *Géographie Universelle*, xiii. 475.

exercise of justice, and their chastity, are things in which they excel all or most nations in the world."[1]

Tachart, Barrow, and Moodie[2] fully confirm Kolben's testimony. Let me only remark that when Kolben wrote that "they are certainly the most friendly, the most liberal and the most benevolent people to one another that ever appeared on the earth" (i. 332), he wrote a sentence which has continually appeared since in the description of savages. When first meeting with primitive races, the Europeans usually make a caricature of their life; but when an intelligent man has stayed among them for a longer time, he generally describes them as the "kindest" or "the gentlest" race on the earth. These very same words have been applied to the Ostyaks, the Samoyedes, the Eskimos, the Dayaks, the Aleoutes, the Papuas, and so on, by the highest authorities. I also remember having read them applied to the Tunguses, the Tchuktchis, the Sioux, and several others. The very frequency of that high commendation already speaks volumes in itself.

The natives of Australia do not stand on a higher level of development than their South African brothers. Their huts are of the same character; very often simple screens are the only protection against cold winds. In their food they are most indifferent: they devour horribly putrefied corpses, and cannibalism is resorted to in times of scarcity. When first discovered by Europeans, they had no implements but in stone or bone, and these were of the roughest description.

[1] P. Kolben, *The Present State of the Cape of Good Hope*, translated from the German by Mr. Medley, London, 1731, vol. i. pp. 59, 71, 333, 336, etc.

[2] Quoted in Waitz's *Anthropologie*, ii. 335 *seq.*

Some tribes had even no canoes, and did not know barter-trade. And yet, when their manners and customs were carefully studied, they proved to be living under that elaborate clan organization which I have mentioned on a preceding page.[1]

The territory they inhabit is usually allotted between the different gentes or clans; but the hunting and fishing territories of each clan are kept in common, and the produce of fishing and hunting belongs to the whole clan; so also the fishing and hunting implements.[2] The meals are taken in common. Like many other savages, they respect certain regulations as to the seasons when certain gums and grasses may be collected.[3] As to their morality altogether, we cannot do better than transcribe the following answers given to the questions of the Paris Anthropological Society by Lumholtz, a missionary who sojourned in North Queensland:[4]—

"The feeling of friendship is known among them; it is strong. Weak people are usually supported; sick people are very well attended to; they never are abandoned or killed. These tribes are cannibals, but they very seldom eat members of their own tribe (when immolated on religious principles, I suppose); they eat strangers only. The parents love their children, play with them, and pet them. Infanticide meets

---

[1] The natives living in the north of Sydney, and speaking the Kamilaroi language, are best known under this aspect, through the capital work of Lorimer Fison and A. W. Howitt, *Kamilaroi and Kurnai*, Melbourne, 1880. See also A. W. Howitt's "Further Note on the Australian Class Systems," in *Journal of the Anthropological Institute*, 1889, vol. xviii. p. 31, showing the wide extension of the same organization in Australia.

[2] *The Folklore, Manners, etc., of Australian Aborigines*, Adelaide, 1879, p. 11.

[3] Grey's *Journals of Two Expeditions of Discovery in North-West and Western Australia*, London, 1841, vol. ii. pp. 237, 298.

[4] *Bulletin de la Société d'Anthropologie*, 1888, vol. xi. p. 652. I abridge the answers.

with common approval. Old people are very well treated, never put to death. No religion, no idols, only a fear of death. Polygamous marriage. Quarrels arising within the tribe are settled by means of duels fought with wooden swords and shields. No slaves; no culture of any kind; no pottery; no dress, save an apron sometimes worn by women. The clan consists of two hundred individuals, divided into four classes of men and four of women; marriage being only permitted within the usual classes, and never within the gens."

For the Papuas, closely akin to the above, we have the testimony of G. L. Bink, who stayed in New Guinea, chiefly in Geelwink Bay, from 1871 to 1883. Here is the essence of his answers to the same questioner:[1]—

"They are sociable and cheerful; they laugh very much. Rather timid than courageous. Friendship is relatively strong among persons belonging to different tribes, and still stronger within the tribe. A friend will often pay the debt of his friend, the stipulation being that the latter will repay it without interest to the children of the lender. They take care of the ill and the old; old people are never abandoned, and in no case are they killed—unless it be a slave who was ill for a long time. War prisoners are sometimes eaten. The children are very much petted and loved. Old and feeble war prisoners are killed, the others are sold as slaves. They have no religion, no gods, no idols, no authority of any description; the oldest man in the family is the judge. In cases of adultery a fine is paid, and part of it goes to the *negoria* (the community). The soil is kept in common, but the crop belongs to those who have grown it. They have pottery, and know barter-trade—the custom being that the merchant gives them the goods, whereupon they return to their houses and bring the native goods required by the merchant; if the latter cannot be obtained, the European goods are returned.[2] They are head-hunters, and in so doing they prosecute blood

---

[1] *Bulletin de la Société d'Anthropologie*, 1888, vol. xi. p. 386.

[2] The same is the practice with the Papuas of Kaimani Bay, who have a high reputation of honesty. "It never happens that the Papua be untrue to his promise," Finsch says in *Neuguinea und seine Bewohner*, Bremen, 1865, p. 829.

revenge. 'Sometimes,' Finsch says, 'the affair is referred to the Rajah of Namototte, who terminates it by imposing a fine.'"

When well treated, the Papuas are very kind. Miklukho-Maclay landed on the eastern coast of New Guinea, followed by one single man, stayed for two years among tribes reported to be cannibals, and left them with regret; he returned again to stay one year more among them, and never had he any conflict to complain of. True that his rule was *never*—under no pretext whatever—to say anything which was not truth, nor make any promise which he could not keep. These poor creatures, who even do not know how to obtain fire, and carefully maintain it in their huts, live under their primitive communism, without any chiefs; and within their villages they have no quarrels worth speaking of. They work in common, just enough to get the food of the day; they rear their children in common; and in the evenings they dress themselves as coquettishly as they can, and dance. Like all savages, they are fond of dancing. Each village has its *barla*, or *balai*—the "long house," "longue maison," or "grande maison"—for the unmarried men, for social gatherings, and for the discussion of common affairs—again a trait which is common to most inhabitants of the Pacific Islands, the Eskimos, the Red Indians, and so on. Whole groups of villages are on friendly terms, and visit each other *en bloc*.

Unhappily, feuds are not uncommon—not in consequence of "overstocking of the area," or "keen competition," and like inventions of a mercantile century, but chiefly in consequence of superstition. As soon as any one falls ill, his friends and relatives come together, and deliberately discuss who might be the cause of the illness. All possible enemies are considered, every one confesses of his own petty quarrels, and finally the real cause is discovered. An enemy from the next village has called it down, and a raid upon that village is decided upon. Therefore, feuds are rather frequent, even between the coast villages, not to say a word of the cannibal mountaineers who are considered as real witches and enemies, though, on a closer acquaintance, they prove to be exactly the same sort of people as their neighbours on the sea-coast.[1]

Many striking pages could be written about the harmony which prevails in the villages of the Polynesian inhabitants of the Pacific Islands. But they belong to a more advanced stage of civilization. So we shall now take our illustrations from the far north. I must mention, however, before leaving the Southern Hemisphere, that even the Fuegians, whose reputation has been so bad, appear under a much better light since they begin to be better known. A few French missionaries who stay among them "know of no act of malevolence to complain of." In their clans, consisting of from 120 to 150 souls, they practise the same primitive communism as the Papuas; they share everything in common, and treat their old people very well. Peace prevails among these tribes.[2]

With the Eskimos and their nearest congeners, the Thlinkets, the Koloshes, and the Aleoutes, we find one of the nearest illustrations of what man may have been during the glacial age. Their implements hardly

[1] *Izvestia* of the Russian Geographical Society, 1880, pp. 161 *seq.* Few books of travel give a better insight into the petty details of the daily life of savages than these scraps from Maklay's note-books.

[2] L. F. Martial, in *Mission Scientifique au Cap Horn*, Paris, 1883, vol. i. pp. 183–201.

differ from those of palæolithic man, and some of their tribes do not yet know fishing: they simply spear the fish with a kind of harpoon.[1] They know the use of iron, but they receive it from the Europeans, or find it on wrecked ships. Their social organization is of a very primitive kind, though they already have emerged from the stage of "communal marriage," even under the gentile restrictions. They live in families, but the family bonds are often broken; husbands and wives are often exchanged.[2] The families, however, remain united in clans, and how could it be otherwise? How could they sustain the hard struggle for life unless by closely combining their forces? So they do, and the tribal bonds are closest where the struggle for life is hardest, namely, in North-East Greenland. The "long house" is their usual dwelling, and several families lodge in it, separated from each other by small partitions of ragged furs, with a common passage in the front. Sometimes the house has the shape of a cross, and in such case a common fire is kept in the centre. The German Expedition which spent a winter close by one of those "long houses" could ascertain that "no quarrel disturbed the peace, no dispute arose about the use of this narrow space" throughout the long winter. "Scolding, or even unkind words, are considered as a misdemeanour, if not produced under the legal form of process, namely, the nith-song."[3] Close cohabitation and close interdependence are

[1] Captain Holm's Expedition to East Greenland.

[2] In Australia whole clans have been seen exchanging all their wives, in order to conjure a calamity (Post, *Studien zur Entwicklungsgeschichte des Familienrechts*, 1890, p. 342). More brotherhood is their specific against calamities.

[3] Dr. H. Rink, *The Eskimo Tribes*, p. 26 (*Meddelelser om Grönland*, vol. xi. 1887).

sufficient for maintaining century after century that deep respect for the interests of the community which is characteristic of Eskimo life. Even in the larger communities of Eskimos, "public opinion formed the real judgment-seat, the general punishment consisting in the offenders being shamed in the eyes of the people."[1]

Eskimo life is based upon communism. What is obtained by hunting and fishing belongs to the clan. But in several tribes, especially in the West, under the influence of the Danes, private property penetrates into their institutions. However, they have an original means for obviating the inconveniences arising from a personal accumulation of wealth which would soon destroy their tribal unity. When a man has grown rich, he convokes the folk of his clan to a great festival, and, after much eating, distributes among them all his fortune. On the Yukon river, Dall saw an Aleoute family distributing in this way ten guns, ten full fur dresses, 200 strings of beads, numerous blankets, ten wolf furs, 200 beavers, and 500 zibelines. After that they took off their festival dresses, gave them away, and, putting on old ragged furs, addressed a few words to their kinsfolk, saying that though they are now poorer than any one of them, they have won their friendship.[2] Like distributions of wealth appear

[1] Dr. Rink, *loc. cit.* p. 24. Europeans, grown in the respect of Roman law, are seldom capable of understanding that force of tribal authority. "In fact," Dr. Rink writes, "it is not the exception, but the rule, that white men who have stayed for ten or twenty years among the Eskimo, return without any real addition to their knowledge of the traditional ideas upon which their social state is based. The white man, whether a missionary or a trader, is firm in his dogmatic opinion that the most vulgar European is better than the most distinguished native."—*The Eskimo Tribes*, p. 31.

[2] Dall, *Alaska and its Resources*, Cambridge, U.S., 1870.

to be a regular habit with the Eskimos, and to take place at a certain season, after an exhibition of all that has been obtained during the year.[1] In my opinion these distributions reveal a very old institution, contemporaneous with the first apparition of personal wealth; they must have been a means for re-establishing equality among the members of the clan, after it had been disturbed by the enrichment of the few. The periodical redistribution of land and the periodical abandonment of all debts which took place in historical times with so many different races (Semites, Aryans, etc.), must have been a survival of that old custom. And the habit of either burying with the dead, or destroying upon his grave, all that belonged to him personally—a habit which we find among all primitive races—must have had the same origin. In fact, while everything that belongs *personally* to the dead is burnt or broken upon his grave, nothing is destroyed of what belonged to him in common with the tribe, such as boats, or the communal implements of fishing. The destruction bears upon personal property alone. At a later epoch this habit becomes a religious ceremony: it receives a mystical interpretation, and is imposed by religion, when public opinion alone proves incapable of enforcing its general observance. And, finally, it is substituted by either burning simple models of the dead man's property (as in China), or by simply carrying his property to the grave and

[1] Dall saw it in Alaska, Jacobsen at Ignitok in the vicinity of the Bering Strait. Gilbert Sproat mentions it among the Vancouver Indians; and Dr. Rink, who describes the periodical exhibitions just mentioned, adds: "The principal use of the accumulation of personal wealth is for *periodically* distributing it." He also mentions (*loc. cit.* p. 31) "the destruction of property for the same purpose" (of maintaining equality).

taking it back to his house after the burial ceremony is over—a habit which still prevails with the Europeans as regards swords, crosses, and other marks of public distinction.[1]

The high standard of the tribal morality of the Eskimos has often been mentioned in general literature. Nevertheless the following remarks upon the manners of the Aleoutes—nearly akin to the Eskimos—will better illustrate savage morality as a whole. They were written, after a ten years' stay among the Aleoutes, by a most remarkable man—the Russian missionary, Veniaminoff. I sum them up, mostly in his own words:—

Endurability (he wrote) is their chief feature. It is simply colossal. Not only do they bathe every morning in the frozen sea, and stand naked on the beach, inhaling the icy wind, but their endurability, even when at hard work on insufficient food, surpasses all that can be imagined. During a protracted scarcity of food, the Aleoute cares first for his children; he gives them all he has, and himself fasts. They are not inclined to stealing; that was remarked even by the first Russian immigrants. Not that they never steal; every Aleoute would confess having sometime stolen something, but it is always a trifle; the whole is so childish. The attachment of the parents to their children is touching, though it is never expressed in words or pettings. The Aleoute is with difficulty moved to make a promise, but once he has made it he will keep it whatever may happen. (An Aleoute made Veniaminoff a gift of dried fish, but it was forgotten on the beach in the hurry of the departure. He took it home. The next occasion to send it to the missionary was in January; and in November and December there was a great scarcity of food in the Aleoute encampment. But the fish was never touched by the starving people, and in January it was sent to its destination.) Their code of morality is both varied and severe. It is considered shameful to be afraid of unavoidable death; to ask pardon from an enemy; to die without ever having killed an enemy; to

[1] See Appendix VIII.

be convicted of stealing; to capsize a boat in the harbour; to be afraid of going to sea in stormy weather; to be the first in a party on a long journey to become an invalid in case of scarcity of food; to show greediness when spoil is divided, in which case every one gives his own part to the greedy man to shame him; to divulge a public secret to his wife; being two persons on a hunting expedition, not to offer the best game to the partner; to boast of his own deeds, especially of invented ones; to scold any one in scorn. Also to beg; to pet his wife in other people's presence, and to dance with her; to bargain personally: selling must always be made through a third person, who settles the price. For a woman it is a shame not to know sewing, dancing and all kinds of woman's work; to pet her husband and children, or even to speak to her husband in the presence of a stranger.[1]

Such is Aleoute morality, which might also be further illustrated by their tales and legends. Let me also add that when Veniaminoff wrote (in 1840) one murder only had been committed since the last century in a population of 60,000 people, and that among 1,800 Aleoutes not one single common law offence had been known for forty years. This will not seem strange if we remark that scolding, scorning, and the use of rough words are absolutely unknown in Aleoute life. Even their children never fight, and never abuse each other in words. All they may say is, "Your mother does not know sewing," or "Your father is blind of one eye."[2]

[1] Veniaminoff, *Memoirs relative to the District of Unalashka* (Russian), 3 vols. St. Petersburg, 1840. Extracts, in English, from the above are given in Dall's *Alaska*. A like description of the Australians' morality is given in *Nature*, xlii. p. 639.

[2] It is most remarkable that several writers (Middendorff, Schrenk, O. Finsch) described the Ostyaks and Samoyedes in almost the same words. Even when drunken, their quarrels are insignificant. "For a hundred years one single murder has been committed in the *tundra*;" "their children never fight;" "anything may be left for years in the tundra, even food and gin, and nobody will touch it;" and so on. Gilbert Sproat "*never* witnessed a fight between two sober natives" of the Aht Indians of Vancouver Island. "Quarrelling is also rare among their children." (Rink, *loc. cit.*) And so on.

Many features of savage life remain, however, a puzzle to Europeans. The high development of tribal solidarity and the good feelings with which primitive folk are animated towards each other, could be illustrated by any amount of reliable testimony. And yet it is not the less certain that those same savages practise infanticide; that in some cases they abandon their old people, and that they blindly obey the rules of blood-revenge. We must then explain the coexistence of facts which, to the European mind, seem so contradictory at the first sight. I have just mentioned how the Aleoute father starves for days and weeks, and gives everything eatable to his child; and how the Bushman mother becomes a slave to follow her child; and I might fill pages with illustrations of the really *tender* relations existing among the savages and their children. Travellers continually mention them incidentally. Here you read about the fond love of a mother; there you see a father wildly running through the forest and carrying upon his shoulders his child bitten by a snake; or a missionary tells you the despair of the parents at the loss of a child whom he had saved, a few years before, from being immolated at its birth; you learn that the "savage" mothers usually nurse their children till the age of four, and that, in the New Hebrides, on the loss of a specially beloved child, its mother, or aunt, will kill herself to take care of it in the other world.[1] And so on.

Like facts are met with by the score; so that, when we see that these same loving parents practise

[1] Gill, quoted in Gerland and Waitz's *Anthropologie*, v. 641. See also pp. 636–640, where many facts of parental and filial love are quoted.

infanticide, we are bound to recognize that the habit (whatever its ulterior transformations may be) took its origin under the sheer pressure of necessity, as an obligation towards the tribe, and a means for rearing the already growing children. The savages, as a rule, do not "multiply without stint," as some English writers put it. On the contrary, they take all kinds of measures for diminishing the birth-rate. A whole series of restrictions, which Europeans certainly would find extravagant, are imposed to that effect, and they are strictly obeyed. But notwithstanding that, primitive folk cannot rear all their children. However, it has been remarked that as soon as they succeed in increasing their regular means of subsistence, they at once begin to abandon the practice of infanticide. On the whole, the parents obey that obligation reluctantly, and as soon as they can afford it they resort to all kinds of compromises to save the lives of their new-born. As has been so well pointed out by my friend Elie Reclus,[1] they invent the lucky and unlucky days of births, and spare the children born on the lucky days; they try to postpone the sentence for a few hours, and then say that if the baby has lived one day it must live all its natural life.[2] They hear the cries of the little ones coming from the forest, and maintain that, if heard, they forbode a misfortune for the tribe; and as they have no baby-farming nor *crèches* for getting rid of the children, every one of them recoils before the necessity of performing the cruel sentence; they prefer to expose the baby in the wood rather than to take its life by violence. Ignorance, not cruelty, maintains infanticide; and, instead of moraliz-

[1] *Primitive Folk*, London, 1891.
[2] Gerland, *loc. cit.* v. 636.

ing the savages with sermons, the missionaries would do better to follow the example of Veniaminoff, who, every year till his old age, crossed the sea of Okhotsk in a miserable boat, or travelled on dogs among his Tchuktchis, supplying them with bread and fishing implements. He thus had really stopped infanticide.

The same is true as regards what superficial observers describe as parricide. We just now saw that the habit of abandoning old people is not so widely spread as some writers have maintained it to be. It has been extremely exaggerated, but it is occasionally met with among nearly all savages; and in such cases it has the same origin as the exposure of children. When a "savage" feels that he is a burden to his tribe; when every morning his share of food is taken from the mouths of the children—and the little ones are not so stoical as their fathers: they cry when they are hungry; when every day he has to be carried across the stony beach, or the virgin forest, on the shoulders of younger people—there are no invalid carriages, nor destitutes to wheel them in savage lands—he begins to repeat what the old Russian peasants say until now-a-day: "*Tchujoi vek zayedayu, Pora na pokoi!*" ("I live other people's life: it is time to retire!") And he retires. He does what the soldier does in a similar case. When the salvation of his detachment depends upon its further advance, and he can move no more, and knows that he must die if left behind, the soldier implores his best friend to render him the last service before leaving the encampment. And the friend, with shivering hands, discharges his gun into the dying body. So the savages do. The old man asks himself to die; he himself insists upon this last duty towards the community, and obtains the consent of the tribe; he digs out his grave; he invites his kinsfolk to the last parting meal. His father has done so, it is now his turn; and he parts with his kinsfolk with marks of affection. The savage so much considers death as part of his *duties* towards his community, that he not only refuses to be rescued (as Moffat has told), but when a woman who had to be immolated on her husband's grave was rescued by missionaries, and was taken to an island, she escaped in the night, crossed a broad sea-arm, swimming and rejoined her tribe, to die on the grave.[1] It has become with them a matter of religion. But the savages, as a rule, are so reluctant to take any one's life otherwise than in fight, that none of them will take upon himself to shed human blood, and they resort to all kinds of stratagems, which have been so falsely interpreted. In most cases, they abandon the old man in the wood, after having given him more than his share of the common food. Arctic expeditions have done the same when they no more could carry their invalid comrades. "Live a few days more! *may be* there will be some unexpected rescue!"

West European men of science, when coming across these facts, are absolutely unable to stand them; they cannot reconcile them with a high development of tribal morality, and they prefer to cast a doubt upon the exactitude of absolutely reliable observers, instead of trying to explain the parallel existence of the two sets of facts: a high tribal morality together with the abandonment of the parents and infanticide. But if these same Europeans were to tell a savage that people, extremely amiable, fond of their own children, and so impressionable that they cry when they see a

[1] Erskine, quoted in Gerland and Waitz's *Anthropologie*, v. 640.

misfortune simulated on the stage, are living in Europe within a stone's throw from dens in which children die from sheer want of food, the savage, too, would not understand them. I remember how vainly I tried to make some of my Tungus friends understand our civilization of individualism: they could not, and they resorted to the most fantastical suggestions. The fact is that a savage, brought up in ideas of a tribal solidarity in everything for bad and for good, is as incapable of understanding a "moral" European, who knows nothing of that solidarity, as the average European is incapable of understanding the savage. But if our scientist had lived amidst a half-starving tribe which does not possess among them all one man's food for so much as a few days to come, he probably might have understood their motives. So also the savage, if he had stayed among us, and received our education, may be, would understand our European indifference towards our neighbours, and our Royal Commissions for the prevention of "baby-farming." "Stone houses make stony hearts," the Russian peasants say. But he ought to live in a stone house first.

Similar remarks must be made as regards cannibalism. Taking into account all the facts which were brought to light during a recent controversy on this subject at the Paris Anthropological Society, and many incidental remarks scattered throughout the "savage" literature, we are bound to recognize that that practice was brought into existence by sheer necessity; but that it was further developed by superstition and religion into the proportions it attained in Fiji or in Mexico. It is a fact that until this day many savages are compelled to devour corpses in the most advanced state of putrefaction, and that in cases of absolute scarcity some of them have had to disinter and to feed upon human corpses, even during an epidemic. These are ascertained facts. But if we now transport ourselves to the conditions which man had to face during the glacial period, in a damp and cold climate, with but little vegetable food at his disposal; if we take into account the terrible ravages which scurvy still makes among underfed natives, and remember that meat and fresh blood are the only restoratives which they know, we must admit that man, who formerly was a granivorous animal, became a flesh-eater during the glacial period. He found plenty of deer at that time, but deer often migrate in the Arctic regions, and sometimes they entirely abandon a territory for a number of years. In such cases his last resources disappeared. During like hard trials, cannibalism has been resorted to even by Europeans, and it was resorted to by the savages. Until the present time, they occasionally devour the corpses of their own dead: they must have devoured then the corpses of those who had to die. Old people died, convinced that by their death they were rendering a last service to the tribe. This is why cannibalism is represented by some savages as of divine origin, as something that has been ordered by a messenger from the sky. But later on it lost its character of necessity, and survived as a superstition. Enemies had to be eaten in order to inherit their courage; and, at a still later epoch, the enemy's eye or heart was eaten for the same purpose; while among other tribes, already having a numerous priesthood and a developed mythology, evil gods, thirsty for human blood, were invented, and human sacrifices required by the priests to appease the gods. In this religious phase of its existence, cannibalism attained its most revolting characters. Mexico is a well-known example; and in Fiji, where the king could eat any one of his subjects, we also find a mighty cast of priests, a complicated theology,[1] and a full development of autocracy. Originated by necessity, cannibalism became, at a later period, a religious institution, and in this form it survived long after it had disappeared from among tribes which certainly practised it in former times, but did not attain the theocratical stage of evolution. The same remark must be made as regards infanticide and the abandonment of parents. In some cases they also have been maintained as a survival of olden times, as a religiously-kept tradition of the past.

I will terminate my remarks by mentioning another custom which also is a source of most erroneous conclusions. I mean the practice of blood-revenge. All savages are under the impression that blood shed must be revenged by blood. If any one has been killed, the murderer must die; if any one has been wounded, the aggressor's blood must be shed. There is no exception to the rule, not even for animals; so the hunter's blood is shed on his return to the village when he has shed the blood of an animal. That is the savages' conception of justice—a conception which yet prevails in Western Europe as regards murder. Now, when both the offender and the offended belong to the same tribe, the tribe and the offended person settle the affair.[2] But when the offender belongs to

[1] W. T. Pritchard, *Polynesian Reminiscences*, London, 1866, p. 363.

[2] It is remarkable, however, that in case of a sentence of death, nobody will take upon himself to be the executioner. Every one

another tribe, and that tribe, for one reason or another, refuses a compensation, then the offended tribe decides to take the revenge itself. Primitive folk so much consider every one's acts as a tribal affair, dependent upon tribal approval, that they easily think the clan responsible for every one's acts. Therefore, the due revenge may be taken upon any member of the offender's clan or relatives.[1] It may often happen, however, that the retaliation goes further than the offence. In trying to inflict a wound, they may kill the offender, or wound him more than they intended to do, and this becomes a cause for a new feud, so that the primitive legislators were careful in requiring the retaliation to be limited to an eye for an eye, a tooth for a tooth, and blood for blood.[2]

It is remarkable, however, that with most primitive folk like feuds are infinitely rarer than might be expected; though with some of them they may attain abnormal proportions, especially with mountaineers

---

throws his stone, or gives his blow with the hatchet, carefully avoiding to give a mortal blow. At a later epoch, the priest will stab the victim with a sacred knife. Still later, it will be the king, until civilization invents the hired hangman. See Bastian's deep remarks upon this subject in *Der Mensch in der Geschichte*, iii. *Die Blutrache*, pp. 1–36. A remainder of this tribal habit, I am told by Professor E. Nys, has survived in military executions till our own times. In the middle portion of the nineteenth century it was the habit to load the rifles of the twelve soldiers called out for shooting the condemned victim, with eleven ball-cartridges and one blank cartridge. As the soldiers never knew who of them had the latter, each one could console his disturbed conscience by thinking that he was not one of the murderers.

[1] In Africa, and elsewhere too, it is a widely-spread habit, that if a theft has been committed, the next clan has to restore the equivalent of the stolen thing, and then look itself for the thief. A. H. Post, *Afrikanische Jurisprudenz*, Leipzig, 1887, vol. i. p. 77.

[2] See Prof. M. Kovalevsky's *Modern Customs and Ancient Law* (Russian), Moscow, 1886, vol. ii., which contains many important considerations upon this subject.

who have been driven to the highlands by foreign invaders, such as the mountaineers of Caucasia, and especially those of Borneo—the Dayaks. With the Dayaks—we were told lately—the feuds had gone so far that a young man could neither marry nor be proclaimed of age before he had secured the head of an enemy. This horrid practice was fully described in a modern English work.[1] It appears, however, that this affirmation was a gross exaggeration. Moreover, Dayak "head-hunting" takes quite another aspect when we learn that the supposed "head-hunter" is not actuated at all by personal passion. He acts under what he considers as a moral obligation towards his tribe, just as the European judge who, in obedience to the same, evidently wrong, principle of "blood for blood," hands over the condemned murderer to the hangman. Both the Dayak and the judge would even feel remorse if sympathy moved them to spare the murderer. That is why the Dayaks, apart from the murders they commit when actuated by their conception of justice, are depicted, by all those who know them, as a most sympathetic people. Thus Carl Bock, the same author who has given such a terrible picture of head-hunting, writes:

"As regards morality, I am bound to assign to the Dayaks a high place in the scale of civilization. . . . Robberies and theft are entirely unknown among them. They also are very truthful. . . . If I did not always get the 'whole truth,' I

[1] See Carl Bock, *The Head-Hunters of Borneo*, London, 1881. I am told, however, by Sir Hugh Law, who was for a long time Governor of Borneo, that the "head-hunting" described in this book is grossly exaggerated. Altogether, my informant speaks of the Dayaks in exactly the same sympathetic terms as Ida Pfeiffer. Let me add that Mary Kingsley speaks in her book on West Africa in the same sympathetic terms of the Fans, who had been represented formerly as the most "terrible cannibals."

always got, at least, nothing but the truth from them. I wish I could say the same of the Malays" (pp. 209 and 210).

Bock's testimony is fully corroborated by that of Ida Pfeiffer. "I fully recognized," she wrote, "that I should be pleased longer to travel among them. I usually found them honest, good, and reserved . . . much more so than any other nation I know."[1] Stoltze used almost the same language when speaking of them. The Dayaks usually have but one wife, and treat her well. They are very sociable, and every morning the whole clan goes out for fishing, hunting, or gardening, in large parties. Their villages consist of big huts, each of which is inhabited by a dozen families, and sometimes by several hundred persons, peacefully living together. They show great respect for their wives, and are fond of their children; and when one of them falls ill, the women nurse him in turn. As a rule they are very moderate in eating and drinking. Such is the Dayak in his real daily life.

It would be a tedious repetition if more illustrations from savage life were given. Wherever we go we find the same sociable manners, the same spirit of solidarity. And when we endeavour to penetrate into the darkness of past ages, we find the same tribal life, the same associations of men, however primitive, for mutual support. Therefore, Darwin was quite right when he saw in man's social qualities the chief factor for his further evolution, and Darwin's vulgarizers are entirely wrong when they maintain the contrary.

> The small strength and speed of man (he wrote), his want of natural weapons, etc., are more than counterbalanced, firstly,

---

[1] Ida Pfeiffer, *Meine zweite Weltreise*, Wien, 1856, vol. i. pp. 116 *seq*. See also Müller and Temminch's *Dutch Possessions in Archipelagic India*, quoted by Elisée Reclus, in *Géographie Universelle*, xiii.

by his intellectual faculties (which, he remarked on another page, have been chiefly or even exclusively gained for the benefit of the community); and secondly, *by his social qualities*, which led him to give and receive aid from his fellow men.[1]

In the last century the "savage" and his "life in the state of nature" were idealized. But now men of science have gone to the opposite extreme, especially since some of them, anxious to prove the animal origin of man, but not conversant with the social aspects of animal life, began to charge the savage with all imaginable "bestial" features. It is evident, however, that this exaggeration is even more unscientific than Rousseau's idealization. The savage is not an ideal of virtue, nor is he an ideal of "savagery." But the primitive man has one quality, elaborated and maintained by the very necessities of his hard struggle for life—he identifies his own existence with that of his tribe; and without that quality mankind never would have attained the level it has attained now.

Primitive folk, as has been already said, so much identify their lives with that of the tribe, that each of their acts, however insignificant, is considered as a tribal affair. Their whole behaviour is regulated by an infinite series of unwritten rules of propriety which are the fruit of their common experience as to what is good or bad—that is, beneficial or harmful for their own tribe. Of course, the reasonings upon which their rules of propriety are based sometimes are absurd in the extreme. Many of them originate in superstition; and altogether, in whatever the savage does, he sees but the immediate consequences of his acts; he cannot foresee their indirect and ulterior con-

[1] *Descent of Man*, second ed., pp. 63, 64.

sequences—thus simply exaggerating a defect with which Bentham reproached civilized legislators. But, absurd or not, the savage obeys the prescriptions of the common law, however inconvenient they may be. He obeys them even more blindly than the civilized man obeys the prescriptions of the written law. His common law is his religion; it is his very habit of living. The idea of the clan is always present to his mind, and self-restriction and self-sacrifice in the interest of the clan are of daily occurrence. If the savage has infringed one of the smaller tribal rules, he is prosecuted by the mockeries of the women. If the infringement is grave, he is tortured day and night by the fear of having called a calamity upon his tribe. If he has wounded by accident any one of his own clan, and thus has committed the greatest of all crimes, he grows quite miserable: he runs away in the woods, and is ready to commit suicide, unless the tribe absolves him by inflicting upon him a physical pain and sheds some of his own blood.[1] Within the tribe everything is shared in common; every morsel of food is divided among all present; and if the savage is alone in the woods, he does not begin eating before he has loudly shouted thrice an invitation to any one who may hear his voice to share his meal.[2]

In short, within the tribe the rule of "each for all" is supreme, so long as the separate family has not yet broken up the tribal unity. But that rule is not extended to the neighbouring clans, or tribes, even when they are federated for mutual protection. Each tribe, or clan, is a separate unity. Just as among mammals

[1] See Bastian's *Mensch in der Geschichte*, iii. p. 7. Also Grey, *loc. cit.* ii. p. 238.

[2] Miklukho-Maclay, *loc. cit.* Same habit with the Hottentots.

and birds, the territory is roughly allotted among separate tribes, and, except in times of war, the boundaries are respected. On entering the territory of his neighbours one must show that he has no bad intentions. The louder one heralds his coming, the more confidence he wins; and if he enters a house, he must deposit his hatchet at the entrance. But no tribe is bound to share its food with the others: it may do so or it may not. Therefore the life of the savage is divided into two sets of actions, and appears under two different ethical aspects: the relations within the tribe, and the relations with the outsiders; and (like our international law) the "inter-tribal" law widely differs from the common law. Therefore, when it comes to a war the most revolting cruelties may be considered as so many claims upon the admiration of the tribe. This double conception of morality passes through the whole evolution of mankind, and maintains itself until now. We Europeans have realized some progress—not immense, at any rate—in eradicating that double conception of ethics; but it also must be said that while we have in some measure extended our ideas of solidarity—in theory, at least—over the nation, and partly over other nations as well, we have lessened the bonds of solidarity within our own nations, and even within our own families.

The appearance of a separate family amidst the clan necessarily disturbs the established unity. A separate family means separate property and accumulation of wealth. We saw how the Eskimos obviate its inconveniences; and it is one of the most interesting studies to follow in the course of ages the different institutions (village communities, guilds, and so on) by means of which the masses endeavoured to maintain the tribal unity, notwithstanding the agencies which were at work to break it down. On the other hand, the first rudiments of knowledge which appeared at an extremely remote epoch, when they confounded themselves with witchcraft, also became a power in the hands of the individual which could be used against the tribe. They were carefully kept in secrecy, and transmitted to the initiated only, in the secret societies of witches, shamans, and priests, which we find among all savages. By the same time, wars and invasions created military authority, as also castes of warriors, whose associations or clubs acquired great powers. However, at no period of man's life were wars the *normal* state of existence. While warriors exterminated each other, and the priests celebrated their massacres, the masses continued to live their daily life, they prosecuted their daily toil. And it is one of the most interesting studies to follow that life of the masses; to study the means by which they maintained their own social organization, which was based upon their own conceptions of equity, mutual aid, and mutual support—of common law, in a word, even when they were submitted to the most ferocious theocracy or autocracy in the State.

CHAPTER IV

MUTUAL AID AMONG THE BARBARIANS

The great migrations.—New organization rendered necessary.—The village community.—Communal work.—Judicial procedure.—Inter-tribal law.—Illustrations from the life of our contemporaries.—Buryates.—Kabyles.—Caucasian mountaineers.—African stems.

It is not possible to study primitive mankind without being deeply impressed by the sociability it has displayed since its very first steps in life. Traces of human societies are found in the relics of both the oldest and the later stone age; and, when we come to observe the savages whose manners of life are still those of neolithic man, we find them closely bound together by an extremely ancient clan organization which enables them to combine their individually weak forces, to enjoy life in common, and to progress. Man is no exception in nature. He also is subject to the great principle of Mutual Aid which grants the best chances of survival to those who best support each other in the struggle for life. These were the conclusions arrived at in the previous chapters.

However, as soon as we come to a higher stage of civilization, and refer to history which already has something to say about that stage, we are bewildered by the struggles and conflicts which it reveals. The old bonds seem entirely to be broken. Stems are seen to fight against stems, tribes against tribes, individuals against individuals; and out of this chaotic contest of hostile forces, mankind issues divided into castes, enslaved to despots, separated into States always ready to wage war against each other. And, with this history of mankind in his hands, the pessimist philosopher triumphantly concludes that warfare and oppression are the very essence of human nature; that the warlike and predatory instincts of man can only be restrained within certain limits by a strong authority which enforces peace and thus gives an opportunity to the few and nobler ones to prepare a better life for humanity in times to come.

And yet, as soon as the every-day life of man during the historical period is submitted to a closer analysis—and so it has been, of late, by many patient students of very early institutions—it appears at once under quite a different aspect. Leaving aside the preconceived ideas of most historians and their pronounced predilection for the dramatic aspects of history, we see that the very documents they habitually peruse are such as to exaggerate the part of human life given to struggles and to underrate its peaceful moods. The bright and sunny days are lost sight of in the gales and storms. Even in our own time, the cumbersome records which we prepare for the future historian, in our Press, our law courts, our Government offices, and even in our fiction and poetry, suffer from the same one-sidedness. They hand down to posterity the most minute descriptions of every war, every battle and skirmish, every contest and act of violence, every kind of individual suffering; but they hardly bear any trace of the countless acts of mutual support and devotion which every one of us knows from his own experience; they hardly take notice of what makes the very essence of our daily life—our social instincts and manners. No wonder, then, if the records of the past were so imperfect. The annalists of old never failed to chronicle the petty wars and calamities which harassed their contemporaries; but they paid no attention whatever to the life of the masses, although the masses chiefly used to toil peacefully while the few indulged in fighting. The epic poems, the inscriptions on monuments, the treaties of peace—nearly all historical documents bear the same character; they deal with breaches of peace, not with peace itself. So that the best-intentioned historian unconsciously draws a distorted picture of the times he endeavours to depict; and, to restore the real proportion between conflict and union, we are now bound to enter into a minute analysis of thousands of small facts and faint indications accidentally preserved in the relics of the past; to interpret them with the aid of comparative ethnology; and, after having heard so much about what used to divide men, to reconstruct stone by stone the institutions which used to unite them.

Ere long history will have to be re-written on new lines, so as to take into account these two currents of human life and to appreciate the part played by each of them in evolution. But in the meantime we may avail ourselves of the immense preparatory work recently done towards restoring the leading features of the second current, so much neglected. From the better-known periods of history we may take some illustrations of the life of the masses, in order to indicate the part played by mutual support during those periods; and, in so doing, we may dispense (for the sake of brevity) from going as far back as the Egyptian, or even the Greek and Roman antiquity. For, in fact, the evolution of mankind has not had the character of one unbroken series. Several times civilization came to an end in one given region, with one given race, and began anew elsewhere, among other races. But at each fresh start it began again with the same clan institutions which we have seen among the savages. So that if we take the last start of our own civilization, when it began afresh in the first centuries of our era, among those whom the Romans called the "barbarians," we shall have the whole scale of evolution, beginning with the gentes and ending in the institutions of our own time. To these illustrations the following pages will be devoted.

Men of science have not yet settled upon the causes which some two thousand years ago drove whole nations from Asia into Europe and resulted in the great migrations of barbarians which put an end to the West Roman Empire. One cause, however, is naturally suggested to the geographer as he contemplates the ruins of populous cities in the deserts of Central Asia, or follows the old beds of rivers now disappeared and the wide outlines of lakes now reduced to the size of mere ponds. It is desiccation: a quite recent desiccation, continued still at a speed which we formerly were not prepared to admit.[1] Against it

[1] Numberless traces of post-pliocene lakes, now disappeared, are found over Central, West, and North Asia. Shells of the same species as those now found in the Caspian Sea are scattered over the surface of the soil as far East as half-way to Lake Aral, and are found in recent deposits as far north as Kazan. Traces of Caspian Gulfs, formerly taken for old beds of the Amu, intersect the Turcoman territory. Deduction must surely be made for temporary, periodical oscillations. But with all that, desiccation is evident, and it progresses at a formerly unexpected speed. Even in the relatively wet parts of South-West Siberia, the succession of reliable surveys, recently

man was powerless. When the inhabitants of North-West Mongolia and East Turkestan saw that water was abandoning them, they had no course open to them but to move down the broad valleys leading to the lowlands, and to thrust westwards the inhabitants of the plains.[1] Stems after stems were thus thrown into Europe, compelling other stems to move and to remove for centuries in succession, westwards and eastwards, in search of new and more or less permanent abodes. Races were mixing with races during those migrations, aborigines with immigrants, Aryans with Ural-Altayans; and it would have been no wonder if the social institutions which had kept them together in their mother-countries had been totally wrecked during the stratification of races which took place in Europe and Asia. But they were *not* wrecked; they simply underwent the modification which was required by the new conditions of life.

The Teutons, the Celts, the Scandinavians, the Slavonians, and others, when they first came in contact with the Romans, were in a transitional state of social organization. The clan unions, based upon a real or supposed common origin, had kept them together for many thousands of years in succession. But these unions could answer their purpose so long only as there were no separate families within the gens or

---

published by Yadrintseff, shows that villages have grown up on what was, eighty years ago, the bottom of one of the lakes of the Tchany group; while the other lakes of the same group, which covered hundreds of square miles some fifty years ago, are now mere ponds. In short, the desiccation of North-West Asia goes on at a rate which must be measured by centuries, instead of by the geological units of time of which we formerly used to speak.

[1] Whole civilizations had thus disappeared, as is proved now by the remarkable discoveries in Mongolia on the Orkhon, and in the depressions of Lukchun (by Dmitri Clements), and of Lob-nor (by Sven Hedin).

clan itself. However, for causes already mentioned, the separate patriarchal family had slowly but steadily developed within the clans, and in the long run it evidently meant the individual accumulation of wealth and power, and the hereditary transmission of both. The frequent migrations of the barbarians and the ensuing wars only hastened the division of the gentes into separate families, while the dispersing of stems and their mingling with strangers offered singular facilities for the ultimate disintegration of those unions which were based upon kinship. The barbarians thus stood in a position of either seeing their clans dissolved into loose aggregations of families, of which the wealthiest, especially if combining sacerdotal functions or military repute with wealth, would have succeeded in imposing their authority upon the others; or of finding out some new form of organization based upon some new principle.

Many stems had no force to resist disintegration: they broke up and were lost for history. But the more vigorous ones did not disintegrate. They came out of the ordeal with a new organization—the *village community*—which kept them together for the next fifteen centuries or more. The conception of a common *territory*, appropriated or protected by common efforts, was elaborated, and it took the place of the vanishing conceptions of common descent. The common gods gradually lost their character of ancestors and were endowed with a local territorial character. They became the gods or saints of a given locality; "the land" was identified with its inhabitants. Territorial unions grew up instead of the consanguine unions of old, and this new organization evidently offered many advantages under the given circumstances. It recognized the independence of the family and even emphasized it, the village community disclaiming all rights of interference in what was going on within the family enclosure; it gave much more freedom to personal initiative; it was not hostile in principle to union between men of different descent, and it maintained at the same time the necessary cohesion of action and thought, while it was strong enough to oppose the dominative tendencies of the minorities of wizards, priests, and professional or distinguished warriors. Consequently it became the primary cell of future organization, and with many nations the village community has retained this character until now.

It is now known, and scarcely contested, that the village community was not a specific feature of the Slavonians, nor even of the ancient Teutons. It prevailed in England during both the Saxon and Norman times, and partially survived till the last century;[1] it was at the bottom of the social organization of old Scotland, old Ireland, and old Wales. In France, the communal possession and the communal allotment of arable land by the village folkmote persisted from the first centuries of our era till the times of Turgot, who found the folkmotes "too noisy" and therefore abolished them. It survived Roman rule in

[1] If I follow the opinions of (to name modern specialists only) Nasse, Kovalevsky, and Vinogradov, and not those of F. Seebohm (Mr. Denman Ross can only be named for the sake of completeness), it is not only because of the deep knowledge and concordance of views of these three writers, but also on account of their perfect knowledge of the village community altogether—a knowledge the want of which is much felt in the otherwise remarkable work of Mr. Seebohm. The same remark applies, in a still higher degree, to the most elegant writings of Fustel de Coulanges, whose opinions and passionate interpretations of old texts are confined to himself.

Italy, and revived after the fall of the Roman Empire. It was the rule with the Scandinavians, the Slavonians, the Finns (in the *pittäyä*, as also, probably, the *kihla-kunta*), the Coures, and the Lives. The village community in India—past and present, Aryan and non-Aryan—is well known through the epoch-making works of Sir Henry Maine; and Elphinstone has described it among the Afghans. We also find it in the Mongolian *oulous*, the Kabyle *thaddart*, the Javanese *dessa*, the Malayan *kota* or *tofa*, and under a variety of names in Abyssinia, the Soudan, in the interior of Africa, with natives of both Americas, with all the small and large tribes of the Pacific archipelagoes. In short, we do not know one single human race or one single nation which has not had its period of village communities. This fact alone disposes of the theory according to which the village community in Europe would have been a servile growth. It is anterior to serfdom, and even servile submission was powerless to break it. It was a universal phase of evolution, a natural outcome of the clan organization, with all those stems, at least, which have played, or play still, some part in history.[1]

[1] The literature of the village community is so vast that but a few works can be named. Those of Sir Henry Maine, F. Seebohm, and Walter's *Das alte Wallis* (Bonn, 1859), are well-known popular sources of information about Scotland, Ireland, and Wales. For France, P. Viollet, *Précis de l'histoire du droit français: Droit privé*, 1886, and several of his monographs in *Bibl. de l'École des Chartes;* Babeau, *Le Village sous l'ancien régime* (the *mir* in the eighteenth century), third edition, 1887; Bonnemère, Doniol, etc. For Italy and Scandinavia, the chief works are named in Laveleye's *Primitive Property*, German version by K. Bücher. For the Finns, Rein's *Föreläsningar*, i. 16; Koskinen, *Finnische Geschichte*, 1874, and various monographs. For the Lives and Coures, Prof. Lutchitzky in *Severnyi Vestnik*, 1891. For the Teutons, besides the well-known works of Maurer, Sohm (*Altdeutsche Reichs- und Gerichts-Verfassung*),

It was a natural growth, and an absolute uniformity in its structure was therefore not possible. As a rule, it was a union between families considered as of common descent and owning a certain territory in common. But with some stems, and under certain circumstances, the families used to grow very numerous before they threw off new buds in the shape of new families; five, six, or seven generations continued to live under the same roof, or within the same enclosure, owning their joint household and cattle in common, and taking their meals at the common hearth. They kept in such case to what ethnology knows as the "joint family," or the "undivided household," which we still see all over China, in India, in the South Slavonian *zadruga*, and occasionally find in Africa, in America, in Denmark, in North Russia, and West France.[1] With other stems, or in other circumstances,

---

also Dahn (*Urzeit*, *Völkerwanderung*, *Langobardische Studien*), Janssen, Wilh. Arnold, etc. For India, besides H. Maine and the works he names, Sir John Phear's *Aryan Village*. For Russia and South Slavonians, see Kavelin, Posnikoff, Sokolovsky, Kovalevsky, Efimenko, Ivanisheff, Klaus, etc. (copious bibliographical index up to 1880 in the *Sbornik svedeniy ob obschinye* of the Russ. Geog. Soc.). For general conclusions, besides Laveleye's *Propriété*, Morgan's *Ancient Society*, Lippert's *Kulturgeschichte*, Post, Dargun, etc., also the lectures of M. Kovalevsky (*Tableau des origines et de l'évolution de la famille et de la propriété*, Stockholm, 1890). Many special monographs ought to be mentioned; their titles may be found in the excellent lists given by P. Viollet in *Droit privé* and *Droit public*. For other races, see subsequent notes.

[1] Several authorities are inclined to consider the joint household as an intermediate stage between the clan and the village community; and there is no doubt that in very many cases village communities have grown up out of undivided families. Nevertheless, I consider the joint household as a fact of a different order. We find it within the gentes; on the other hand, we cannot affirm that joint families have existed at any period without belonging either to a gens or to a village community, or to a *Gau*. I conceive the early village communities as slowly originating directly from the gentes, and consisting, according to racial and local circumstances, either of several joint

not yet well specified, the families did not attain the same proportions ; the grandsons, and occasionally the sons, left the household as soon as they were married, and each of them started a new cell of his own. But, joint or not, clustered together or scattered in the woods, the families remained united into village communities ; several villages were grouped into tribes ; and the tribes joined into confederations. Such was the social organization which developed among the so-called "barbarians," when they began to settle more or less permanently in Europe.

A very long evolution was required before the gentes, or clans, recognized the separate existence of a patriarchal family in a separate hut ; but even after that had been recognized, the clan, as a rule, knew no personal inheritance of property. The few things which might have belonged personally to the individual were either destroyed on his grave or buried with him. The village community, on the contrary, fully recognized the private accumulation of wealth within the family and its hereditary transmission. But wealth was conceived exclusively in the shape of *movable* property, including cattle, implements, arms, and the dwelling-house which—"like all things that can be destroyed by fire"—belonged to the same category.[1] As to private property in land, the village community did not, and could not, recognize anything of the kind,

---

families, or of both joint and simple families, or (especially in the case of new settlements) of simple families only. If this view be correct, we should not have the right of establishing the series : gens, compound family, village community—the second member of the series having not the same ethnological value as the two others. See Appendix IX.

[1] Stobbe, *Beiträge zur Geschichte des deutschen Rechtes*, p. 62.

and, as a rule, it does not recognize it now. The land was the common property of the tribe, or of the whole stem, and the village community itself owned its part of the tribal territory so long only as the tribe did not claim a re-distribution of the village allotments. The clearing of the woods and the breaking of the prairies being mostly done by the communities or, at least, by the joint work of several families—always with the consent of the community—the cleared plots were held by each family for a term of four, twelve, or twenty years, after which term they were treated as parts of the arable land owned in common. Private property, or possession "for ever," was as incompatible with the very principles and the religious conceptions of the village community as it was with the principles of the gens; so that a long influence of the Roman law and the Christian Church, which soon accepted the Roman principles, were required to accustom the barbarians to the idea of private property in land being possible.[1] And yet, even when such property, or possession for an unlimited time, was recognized, the owner of a separate estate remained a co-proprietor in the waste lands, forests, and grazing-grounds. Moreover, we continually see, especially in the history of Russia, that when a few families, acting separately, had taken possession of some land belonging to tribes which were treated as strangers, they very soon united together, and constituted a village community which

[1] The few traces of private property in land which are met with in the early barbarian period are found with such stems (the Batavians, the Franks in Gaul) as have been for a time under the influence of Imperial Rome. See Inama-Sternegg's *Die Ausbildung der grossen Grundherrschaften in Deutschland*, Bd. i. 1878. Also, Besseler, *Neubruch nach dem älteren deutschen Recht*, pp. 11–12, quoted by Kovalevsky, *Modern Custom and Ancient Law*, Moscow, 1886, i. 134.

in the third or fourth generation began to profess a community of origin.

A whole series of institutions, partly inherited from the clan period, have developed from that basis of common ownership of land during the long succession of centuries which was required to bring the barbarians under the dominion of States organized upon the Roman or Byzantine pattern. The village community was not only a union for guaranteeing to each one his fair share in the common land, but also a union for common culture, for mutual support in all possible forms, for protection from violence, and for a further development of knowledge, national bonds, and moral conceptions; and every change in the judicial, military, educational, or economical manners had to be decided at the folkmotes of the village, the tribe, or the confederation. The community being a continuation of the gens, it inherited all its functions. It was the *universitas*, the *mir*—a world in itself.

Common hunting, common fishing, and common culture of the orchards or the plantations of fruit trees was the rule with the old gentes. Common agriculture became the rule in the barbarian village communities. True, that direct testimony to this effect is scarce, and in the literature of antiquity we only have the passages of Diodorus and Julius Cæsar relating to the inhabitants of the Lipari Islands, one of the Celt-Iberian tribes, and the Sueves. But there is no lack of evidence to prove that common agriculture was practised among some Teuton tribes, the Franks, and the old Scotch, Irish, and Welsh.[1] As to the later survivals of the

[1] Maurer's *Markgenossenschaft;* Lamprecht's "Wirthschaft und Recht der Franken zur Zeit der Volksrechte," in *Histor. Taschenbuch*, 1883; Seebohm's *The English Village Community*, ch. vi., vii., and ix.

same practice, they simply are countless. Even in perfectly Romanized France, common culture was habitual some five and twenty years ago in the Morbihan (Brittany).[1] The old Welsh *cyvar*, or joint team, as well as the common culture of the land allotted to the use of the village sanctuary are quite common among the tribes of Caucasus the least touched by civilization,[2] and like facts are of daily occurrence among the Russian peasants. Moreover, it is well known that many tribes of Brazil, Central America, and Mexico used to cultivate their fields in common, and that the same habit is widely spread among some Malayans, in New Caledonia, with several Negro stems, and so on.[3] In short, communal culture is so habitual with many Aryan, Ural-Altayan, Mongolian, Negro, Red Indian, Malayan, and Melanesian stems that we must consider it as a universal—though not as the only possible—form of primitive agriculture.[4]

Communal cultivation does not, however, imply by necessity communal consumption. Already under the clan organization we often see that when the boats laden with fruits or fish return to the village, the food they bring in is divided among the huts and the "long

[1] Letourneau, in *Bulletin de la Soc. d'Anthropologie*, 1888, vol. xi. p. 476.

[2] Walter, *Das alte Wallis*, p. 323; Dm. Bakradze and N. Khoudadoff in Russian *Zapiski* of the Caucasian Geogr. Society, xiv. Part I.

[3] Bancroft's *Native Races*; Waitz, *Anthropologie*, iii. 423; Montrozier, in *Bull. Soc. d'Anthropologie*, 1870; Post's *Studien*, etc.

[4] A number of works, by Ory, Luro, Laudes, and Sylvestre, on the village community in Annam, proving that it has had there the same forms as in Germany or Russia, is mentioned in a review of these works by Jobbé-Duval, in *Nouvelle Revue historique de droit français et étranger*, October and December, 1896. A good study of the village community of Peru, before the establishment of the power of the Incas, has been brought out by Heinrich Cunow (*Die Soziale Verfassung des Inka-Reichs*, Stuttgart, 1896. The communal possession of land and communal culture are described in that work.

houses" inhabited by either several families or the youth, and is cooked separately at each separate hearth. The habit of taking meals in a narrower circle of relatives or associates thus prevails at an early period of clan life. It became the rule in the village community. Even the food grown in common was usually divided between the households after part of it had been laid in store for communal use. However, the tradition of communal meals was piously kept alive; every available opportunity, such as the commemoration of the ancestors, the religious festivals, the beginning and the end of field work, the births, the marriages, and the funerals, being seized upon to bring the community to a common meal. Even now this habit, well known in this country as the "harvest supper," is the last to disappear. On the other hand, even when the fields had long since ceased to be tilled and sown in common, a variety of agricultural work continued, and continues still, to be performed by the community. Some part of the communal land is still cultivated in many cases in common, either for the use of the destitute, or for refilling the communal stores, or for using the produce at the religious festivals. The irrigation canals are digged and repaired in common. The communal meadows are mown by the community; and the sight of a Russian commune mowing a meadow—the men rivalling each other in their advance with the scythe, while the women turn the grass over and throw it up into heaps—is one of the most inspiring sights; it shows what human work might be and ought to be. The hay, in such case, is divided among the separate households, and it is evident that no one has the right of taking hay from a neighbour's stack without his permission; but the limitation of this last rule among the Caucasian Ossetes is most noteworthy. When the cuckoo cries and announces that spring is coming, and that the meadows will soon be clothed again with grass, every one in need has the right of taking from a neighbour's stack the hay he wants for his cattle.[1] The old communal rights are thus re-asserted, as if to prove how contrary unbridled individualism is to human nature.

When the European traveller lands in some small island of the Pacific, and, seeing at a distance a grove of palm trees, walks in that direction, he is astonished to discover that the little villages are connected by roads paved with big stones, quite comfortable for the unshod natives, and very similar to the "old roads" of the Swiss mountains. Such roads were traced by the "barbarians" all over Europe, and one must have travelled in wild, thinly-peopled countries, far away from the chief lines of communication, to realize in full the immense work that must have been performed by the barbarian communities in order to conquer the woody and marshy wilderness which Europe was some two thousand years ago. Isolated families, having no tools, and weak as they were, could not have conquered it; the wilderness would have overpowered them. Village communities alone, working in common, could master the wild forests, the sinking marshes, and the endless steppes. The rough roads, the ferries, the wooden bridges taken away in the winter and rebuilt after the spring flood was over, the fences and the palisaded walls of the villages, the earthen forts and the small towers with which the territory was dotted—all these were the work of the barbarian communities. And when a community grew numerous it used to

[1] Kovalevsky, *Modern Custom and Ancient Law*, i. 115.

throw off a new bud. A new community arose at a distance, thus step by step bringing the woods and the steppes under the dominion of man. The whole making of European nations was such a budding of the village communities. Even now-a-days the Russian peasants, if they are not quite broken down by misery, migrate in communities, and they till the soil and build the houses in common when they settle on the banks of the Amur, or in Manitoba. And even the English, when they first began to colonize America, used to return to the old system; they grouped into village communities.[1]

The village community was the chief arm of the barbarians in their hard struggle against a hostile nature. It also was the bond they opposed to oppression by the cunningest and the strongest which so easily might have developed during those disturbed times. The imaginary barbarian—the man who fights and kills at his mere caprice—existed no more than the "bloodthirsty" savage. The real barbarian was living, on the contrary, under a wide series of institutions, imbued with considerations as to what may be useful or noxious to his tribe or confederation, and these institutions were piously handed down from generation to generation in verses and songs, in proverbs or triads, in sentences and instructions. The more we study them the more we recognize the narrow bonds which united men in their villages. Every quarrel arising between two individuals was treated as a communal affair—even the offensive words that might have been uttered during a quarrel being considered as an offence

[1] Palfrey, *History of New England*, ii. 13; quoted in Maine's *Village Communities*, New York, 1876, p. 201.

to the community and its ancestors. They had to be repaired by amends made both to the individual and the community;[1] and if a quarrel ended in a fight and wounds, the man who stood by and did not interpose was treated as if he himself had inflicted the wounds.[2]

The judicial procedure was imbued with the same spirit. Every dispute was brought first before mediators or arbiters, and it mostly ended with them, the arbiters playing a very important part in barbarian society. But if the case was too grave to be settled in this way, it came before the folkmote, which was bound "to find the sentence," and pronounced it in a conditional form; that is, "such compensation was due, if the wrong be proved," and the wrong had to be proved or disclaimed by six or twelve persons confirming or denying the fact by oath; ordeal being resorted to in case of contradiction between the two sets of jurors. Such procedure, which remained in force for more than two thousand years in succession, speaks volumes for itself; it shows how close were the bonds between all members of the community. Moreover, there was no other authority to enforce the decisions of the folkmote besides its own moral authority. The only possible menace was that the community might declare the rebel an outlaw, but even this menace was reciprocal. A man discontented with the folkmote could declare that he would abandon the tribe and go over to another tribe—a most dreadful menace, as it was sure to bring all kinds of misfortunes upon a tribe that might have been unfair to one of its

[1] Königswarter, *Études sur le développement des sociétés humaines*, Paris, 1850.

[2] This is, at least, the law of the Kalmucks, whose customary law bears the closest resemblance to the laws of the Teutons, the old Slavonians, etc.

members.[1] A rebellion against a right decision of the customary law was simply "inconceivable," as Henry Maine has so well said, because "law, morality, and fact" could not be separated from each other in those times.[2] The moral authority of the commune was so great that even at a much later epoch, when the village communities fell into submission to the feudal lord, they maintained their judicial powers; they only permitted the lord, or his deputy, to "find" the above conditional sentence in accordance with the customary law he had sworn to follow, and to levy for himself the fine (the *fred*) due to the commune. But for a long time, the lord himself, if he remained a co-proprietor in the waste land of the commune, submitted in communal affairs to its decisions. Noble or ecclesiastic, he had to submit to the folkmote—*Wer daselbst Wasser und Weid genusst, muss gehorsam sein*—"Who enjoys here the right of water and pasture must obey"—was the old saying. Even when the peasants became serfs under the lord, he was bound to appear before the folkmote when they summoned him.[3]

In their conceptions of justice the barbarians evidently did not much differ from the savages. They also maintained the idea that a murder must be followed by putting the murderer to death; that wounds had to be punished by equal wounds, and that the wronged family was bound to fulfil the sentence of the customary

[1] The habit is in force still with many African and other tribes.

[2] *Village Communities*, pp. 65–68 and 199.

[3] Maurer (*Gesch. der Markverfassung*, § 29, 97) is quite decisive upon this subject. He maintains that "All members of the community . . . . the laic and clerical lords as well, often also the partial co-possessors (*Markberechtigte*), and even strangers to the Mark, were submitted to its jurisdiction" (p. 312). This conception remained locally in force up to the fifteenth century.

law. This was a holy duty, a duty towards the ancestors, which had to be accomplished in broad daylight, never in secrecy, and rendered widely known. Therefore the most inspired passages of the sagas and epic poetry altogether are those which glorify what was supposed to be justice. The gods themselves joined in aiding it. However, the predominant feature of barbarian justice is, on the one hand, to limit the numbers of persons who may be involved in a feud, and, on the other hand, to extirpate the brutal idea of blood for blood and wounds for wounds, by substituting for it the system of compensation. The barbarian codes—which were collections of common law rules written down for the use of judges—"first permitted, then encouraged, and at last enforced," compensation instead of revenge.[1] The compensation has, however, been totally misunderstood by those who represented it as a fine, and as a sort of *carte blanche* given to the rich man to do whatever he liked. The compensation money (*wergeld*), which was quite different from the fine or *fred*,[2] was habitually so high for all kinds of active offences that it certainly was no encouragement for such offences. In case of a murder it usually exceeded all the possible fortune of the murderer. "Eighteen times eighteen cows" is the compensation with the Ossetes who do not know how to reckon above eighteen, while with the African tribes it attains 800 cows or 100 camels with their young, or 416 sheep

[1] Königswarter, *loc. cit.* p. 50; J. Thrupp, *Historical Law Tracts*, London, 1843, p. 106.

[2] Königswarter has shown that the *fred* originated from an offering which had to be made to appease the ancestors. Later on, it was paid to the community, for the breach of peace; and still later to the judge, or king, or lord, when they had appropriated to themselves the rights of the community.

in the poorer tribes.[1] In the great majority of cases, the compensation money could not be paid at all, so that the murderer had no issue but to induce the wronged family, by repentance, to adopt him. Even now, in the Caucasus, when feuds come to an end, the offender touches with his lips the breast of the oldest woman of the tribe, and becomes a "milk-brother" to all men of the wronged family.[2] With several African tribes he must give his daughter, or sister, in marriage to some one of the family; with other tribes he is bound to marry the woman whom he has made a widow; and in all cases he becomes a member of the family, whose opinion is taken in all important family matters.[3]

Far from acting with disregard to human life, the barbarians, moreover, knew nothing of the horrid punishments introduced at a later epoch by the laic and canonic laws under Roman and Byzantine influence. For, if the Saxon code admitted the death penalty rather freely, even in cases of incendiarism and armed robbery, the other barbarian codes pronounced it exclusively in cases of betrayal of one's kin, and sacrilege against the community's gods, as the only means to appease the gods.

All this, as seen, is very far from the supposed

[1] Post's *Bausteine* and *Afrikanische Jurisprudenz*, Oldenburg, 1887, vol. i. pp. 64 *seq.*; Kovalevsky, *loc. cit.* ii. 164–189.

[2] O. Miller and M. Kovalevsky, "In the Mountaineer Communities of Kabardia," in *Vestnik Evropy*, April, 1884. With the Shakhsevens of the Mugan Steppe, blood feuds always end by marriage between the two hostile sides (Markoff, in appendix to the *Zapiski* of the Caucasian Geogr. Soc., xiv. 1, 21).

[3] Post, in *Afrik. Jurisprudenz*, gives a series of facts illustrating the conceptions of equity inrooted among the African barbarians. The same may be said of all serious examinations into barbarian common law.

"moral dissoluteness" of the barbarians. On the contrary, we cannot but admire the deeply moral principles elaborated within the early village communities which found their expression in Welsh triads, in legends about King Arthur, in Brehon commentaries,[1] in old German legends and so on, or find still their expression in the sayings of the modern barbarians. In his introduction to *The Story of Burnt Njal*, George Dasent very justly sums up as follows the qualities of a Northman, as they appear in the sagas:—

> To do what lay before him openly and like a man, without fear of either foes, fiends, or fate; . . . to be free and daring in all his deeds; to be gentle and generous to his friends and kinsmen; to be stern and grim to his foes [those who are under the *lex talionis*], but even towards them to fulfil all bounden duties. . . . To be no truce-breaker, nor tale-bearer, nor backbiter. To utter nothing against any man that he would not dare to tell him to his face. To turn no man from his door who sought food or shelter, even though he were a foe.[2]

The same or still better principles permeate the Welsh epic poetry and triads. To act "according to the nature of mildness and the principles of equity," without regard to the foes or to the friends, and "to repair the wrong," are the highest duties of man; "evil is death, good is life," exclaims the poet legislator.[3] "The World would be fool, if agreements made on lips were not honourable"—the Brehon law says. And the humble Shamanist Mordovian, after having praised the same qualities, will add, moreover, in his principles of customary law, that "among neighbours

[1] See the excellent chapter, "Le droit de la Vieille Irlande," (also "Le Haut Nord") in *Études de droit international et de droit politique*, by Prof. E. Nys, Bruxelles, 1896.

[2] Introduction, p. xxxv.

[3] *Das alte Wallis*, pp. 343–350.

the cow and the milking-jar are in common;" that "the cow must be milked for yourself and him who may ask milk;" that "the body of a child reddens from the stroke, but the face of him who strikes reddens from shame;"[1] and so on. Many pages might be filled with like principles expressed and followed by the "barbarians."

One feature more of the old village communities deserves a special mention. It is the gradual extension of the circle of men embraced by the feelings of solidarity. Not only the tribes federated into stems, but the stems as well, even though of different origin, joined together in confederations. Some unions were so close that, for instance, the Vandals, after part of their confederation had left for the Rhine, and thence went over to Spain and Africa, respected for forty consecutive years the landmarks and the abandoned villages of their confederates, and did not take possession of them until they had ascertained through envoys that their confederates did not intend to return. With other barbarians, the soil was cultivated by one part of the stem, while the other part fought on or beyond the frontiers of the common territory. As to the leagues between several stems, they were quite habitual. The Sicambers united with the Cherusques and the Sueves, the Quades with the Sarmates; the Sarmates with the Alans, the Carpes, and the Huns. Later on, we also see the conception of nations gradually developing in Europe, long before anything like a State had grown in any part of the continent occupied by the barbarians. These nations—for it is impossible

[1] Maynoff, "Sketches of the Judicial Practices of the Mordovians," in the ethnographical *Zapiski* of the Russian Geographical Society, 1885, pp. 236, 257.

to refuse the name of a nation to the Merovingian France, or to the Russia of the eleventh and twelfth century—were nevertheless kept together by nothing else but a community of language, and a tacit agreement of the small republics to take their dukes from none but one special family.

Wars were certainly unavoidable; migration means war; but Sir Henry Maine has already fully proved in his remarkable study of the tribal origin of International Law, that "Man has never been so ferocious or so stupid as to submit to such an evil as war without some kind of effort to prevent it," and he has shown how exceedingly great is "the number of ancient institutions which bear the marks of a design to stand in the way of war, or to provide an alternative to it."[1] In reality, man is so far from the warlike being he is supposed to be, that when the barbarians had once settled they so rapidly lost the very habits of warfare that very soon they were compelled to keep special dukes followed by special *scholæ* or bands of warriors, in order to protect them from possible intruders. They preferred peaceful toil to war, the very peacefulness of man being the cause of the specialization of the warrior's trade, which specialization resulted later on in serfdom and in all the wars of the "States period" of human history.

History finds great difficulties in restoring to life the institutions of the barbarians. At every step the historian meets with some faint indication which he is unable to explain with the aid of his own documents only. But a broad light is thrown on the past as soon

[1] Henry Maine, *International Law*, London, 1888, pp. 11–13. E. Nys, *Les origines du droit international*, Bruxelles, 1894.

as we refer to the institutions of the very numerous tribes which are still living under a social organization almost identical with that of our barbarian ancestors. Here we simply have the difficulty of choice, because the islands of the Pacific, the steppes of Asia, and the tablelands of Africa are real historical museums containing specimens of all possible intermediate stages which mankind has lived through, when passing from the savage gentes up to the States' organization. Let us, then, examine a few of those specimens.

If we take the village communities of the Mongol Buryates, especially those of the Kudinsk Steppe on the upper Lena which have better escaped Russian influence, we have fair representatives of barbarians in a transitional state, between cattle-breeding and agriculture.[1] These Buryates are still living in "joint families;" that is, although each son, when he is married, goes to live in a separate hut, the huts of at least three generations remain within the same enclosure, and the joint family work in common in their fields, and own in common their joint households and their cattle, as well as their "calves' grounds" (small fenced patches of soil kept under soft grass for the rearing of calves). As a rule, the meals are taken separately in each hut; but when meat is roasted, all the twenty to sixty members of the joint household feast together. Several joint households which live in a cluster, as well as several smaller families settled in the same village—mostly *débris* of joint households accidentally broken up—make the *oulous*, or the village

[1] A Russian historian, the Kazan Professor Schapoff, who was exiled in 1862 to Siberia, has given a good description of their institutions in the *Izvestia* of the East-Siberian Geographical Society, vol. v. 1874.

community; several *oulouses* make a tribe; and the forty-six tribes, or clans, of the Kudinsk Steppe are united into one confederation. Smaller and closer confederations are entered into, as necessity arises for special wants, by several tribes. They know no private property in land—the land being held in common by the *oulous*, or rather by the confederation, and if it becomes necessary, the territory is re-allotted between the different *oulouses* at a folkmote of the tribe, and between the forty-six tribes at a folkmote of the confederation. It is worthy of note that the same organization prevails among all the 250,000 Buryates of East Siberia, although they have been for three centuries under Russian rule, and are well acquainted with Russian institutions.

With all that, inequalities of fortune rapidly develop among the Buryates, especially since the Russian Government is giving an exaggerated importance to their elected *taishas* (princes), whom it considers as responsible tax-collectors and representatives of the confederations in their administrative and even commercial relations with the Russians. The channels for the enrichment of the few are thus many, while the impoverishment of the great number goes hand in hand, through the appropriation of the Buryate lands by the Russians. But it is a habit with the Buryates, especially those of Kudinsk—and habit is more than law—that if a family has lost its cattle, the richer families give it some cows and horses that it may recover. As to the destitute man who has no family, he takes his meals in the huts of his congeners; he enters a hut, takes—by right, not for charity—his seat by the fire, and shares the meal which always is scrupulously divided into equal parts; he sleeps where he has taken his evening meal. Altogether, the Russian conquerors of Siberia were so much struck by the communistic practices of the Buryates, that they gave them the name of *Bratskiye*—"the Brotherly Ones"—and reported to Moscow: "With them everything is in common; whatever they have is shared in common." Even now, when the Lena Buryates sell their wheat, or send some of their cattle to be sold to a Russian butcher, the families of the *oulous*, or the tribe, put their wheat and cattle together, and sell it as a whole. Each *oulous* has, moreover, its grain store for loans in case of need, its communal baking oven (the *four banal* of the old French communities), and its blacksmith, who, like the blacksmith of the Indian communities,[1] being a member of the community, is never paid for his work within the community. He must make it for nothing, and if he utilizes his spare time for fabricating the small plates of chiselled and silvered iron which are used in Buryate land for the decoration of dress, he may occasionally sell them to a woman from another clan, but to the women of his own clan the attire is presented as a gift. Selling and buying cannot take place within the community, and the rule is so severe that when a richer family hires a labourer the labourer must be taken from another clan or from among the Russians. This habit is evidently not specific to the Buryates; it is so widely spread among the modern barbarians, Aryan and Ural-Altayan, that it must have been universal among our ancestors.

The feeling of union within the confederation is kept alive by the common interests of the tribes, their folkmotes, and the festivities which are usually kept in

[1] Sir Henry Maine's *Village Communities*, New York, 1876, pp. 193–196.

connection with the folkmotes. The same feeling is, however, maintained by another institution, the *aba*, or common hunt, which is a reminiscence of a very remote past. Every autumn, the forty-six clans of Kudinsk come together for such a hunt, the produce of which is divided among all the families. Moreover, national *abas*, to assert the unity of the whole Buryate nation, are convoked from time to time. In such cases, all Buryate clans which are scattered for hundreds of miles west and east of Lake Baikal, are bound to send their delegate hunters. Thousands of men come together, each one bringing provisions for a whole month. Every one's share must be equal to all the others, and therefore, before being put together, they are weighed by an elected elder (always "with the hand": scales would be a profanation of the old custom). After that the hunters divide into bands of twenty, and the parties go hunting according to a well-settled plan. In such *abas* the entire Buryate nation revives its epic traditions of a time when it was united in a powerful league. Let me add that such communal hunts are quite usual with the Red Indians and the Chinese on the banks of the Usuri (the *kada*).[1]

With the Kabyles, whose manners of life have been so well described by two French explorers,[2] we have barbarians still more advanced in agriculture. Their fields, irrigated and manured, are well attended to, and in the hilly tracts every available plot of land is cultivated by the spade. The Kabyles have known many vicissitudes in their history; they have followed for

[1] Nazaroff, *The North Usuri Territory* (Russian), St. Petersburg, 1887, p. 65.

[2] Hanoteau et Letourneux, *La Kabylie*, 3 vols. Paris, 1883.

some time the Mussulman law of inheritance, but, being adverse to it, they have returned, 150 years ago, to the tribal customary law of old. Accordingly, their land-tenure is of a mixed character, and private property in land exists side by side with communal possession. Still, the basis of their present organization is the village community, the *thaddart*, which usually consists of several joint families (*kharoubas*), claiming a community of origin, as well as of smaller families of strangers. Several villages are grouped into clans or tribes (*ârch*); several tribes make the confederation (*thak'ebilt*); and several confederations may occasionally enter into a league, chiefly for purposes of armed defence.

The Kabyles know no authority whatever besides that of the *djemmâa*, or folkmote of the village community. All men of age take part in it, in the open air, or in a special building provided with stone seats, and the decisions of the *djemmâa* are evidently taken at unanimity: that is, the discussions continue until all present agree to accept, or to submit to, some decision. There being no authority in a village community to impose a decision, this system has been practised by mankind wherever there have been village communities, and it is practised still wherever they continue to exist, *i. e.* by several hundred million men all over the world. The *djemmâa* nominates its executive—the elder, the scribe, and the treasurer; it assesses its own taxes; and it manages the repartition of the common lands, as well as all kinds of works of public utility. A great deal of work is done in common: the roads, the mosques, the fountains, the irrigation canals, the towers erected for protection from robbers, the fences, and so on, are built by the village community; while the high-roads, the larger mosques, and the great market-places are the work of the tribe. Many traces of common culture continue to exist, and the houses continue to be built by, or with the aid of, all men and women of the village. Altogether, the "aids" are of daily occurrence, and are continually called in for the cultivation of the fields, for harvesting, and so on. As to the skilled work, each community has its blacksmith, who enjoys his part of the communal land, and works for the community; when the tilling season approaches he visits every house, and repairs the tools and the ploughs, without expecting any pay, while the making of new ploughs is considered as a pious work which can by no means be recompensed in money, or by any other form of salary.

As the Kabyles already have private property, they evidently have both rich and poor among them. But like all people who closely live together, and know how poverty begins, they consider it as an accident which may visit every one. "Don't say that you will never wear the beggar's bag, nor go to prison," is a proverb of the Russian peasants; the Kabyles practise it, and no difference can be detected in the external behaviour between rich and poor; when the poor convokes an "aid," the rich man works in his field, just as the poor man does it reciprocally in his turn.[1] Moreover, the *djemmâas* set aside certain gardens and fields, sometimes cultivated in common, for the use of

[1] To convoke an "aid, or "bee," some kind of meal must be offered to the community. I am told by a Caucasian friend that in Georgia, when the poor man wants an "aid," he borrows from the rich man a sheep or two to prepare the meal, and the community bring, in addition to their work, so many provisions that he may repay the debt. A similar habit exists with the Mordovians.

the poorest members. Many like customs continue to exist. As the poorer families would not be able to buy meat, meat is regularly bought with the money of the fines, or the gifts to the *djemmâa*, or the payments for the use of the communal olive-oil basins, and it is distributed in equal parts among those who cannot afford buying meat themselves. And when a sheep or a bullock is killed by a family for its own use on a day which is not a market day, the fact is announced in the streets by the village crier, in order that sick people and pregnant women may take of it what they want. Mutual support permeates the life of the Kabyles, and if one of them, during a journey abroad, meets with another Kabyle in need, he is bound to come to his aid, even at the risk of his own fortune and life; if this has not been done, the *djemmâa* of the man who has suffered from such neglect may lodge a complaint, and the *djemmâa* of the selfish man will at once make good the loss. We thus come across a custom which is familiar to the students of the mediæval merchant guilds. Every stranger who enters a Kabyle village has right to housing in the winter, and his horses can always graze on the communal lands for twenty-four hours. But in case of need he can reckon upon an almost unlimited support. Thus, during the famine of 1867-68, the Kabyles received and fed every one who sought refuge in their villages, without distinction of origin. In the district of Dellys, no less than 12,000 people who came from all parts of Algeria, and even from Morocco, were fed in this way. While people died from starvation all over Algeria, there was not one single case of death due to this cause on Kabylian soil. The *djemmâas*, depriving themselves of necessaries, organized relief, without ever asking any aid from the Government, or uttering the slightest complaint; they considered it as a natural duty. And while among the European settlers all kind of police measures were taken to prevent thefts and disorder resulting from such an influx of strangers, nothing of the kind was required on the Kabyles' territory: the *djemmâas* needed neither aid nor protection from without.[1]

I can only cursorily mention two other most interesting features of Kabyle life; namely, the *anaya*, or protection granted to wells, canals, mosques, market-places, some roads, and so on, in case of war, and the *çofs*. In the *anaya* we have a series of institutions both for diminishing the evils of war and for preventing conflicts. Thus the market-place is *anaya*, especially if it stands on a frontier and brings Kabyles and strangers together; no one dares disturb peace in the market, and if a disturbance arises, it is quelled at once by the strangers who have gathered in the market town. The road upon which the women go from the village to the fountain also is *anaya* in case of war; and so on. As to the *çof*, it is a widely-spread form of association, having some characters of the mediæval *Bürgschaften* or *Gegilden*, as well as of societies both for mutual protection and for various purposes—intellectual, political, and emotional—which cannot be satisfied by the territorial organization of the village, the clan, and the confederation. The *çof* knows no territorial limits; it recruits its members in various villages, even among strangers; and it pro-

[1] Hanoteau et Letourneux, *La Kabylie*, ii. 58. The same respect to strangers is the rule with the Mongols. The Mongol who has refused his roof to a stranger pays the full blood-compensation if the stranger has suffered therefrom (Bastian, *Der Mensch in der Geschichte*, iii. 231).

tects them in all possible eventualities of life. Altogether, it is an attempt at supplementing the territorial grouping by an extra-territorial grouping intended to give an expression to mutual affinities of all kinds across the frontiers. The free international association of individual tastes and ideas, which we consider as one of the best features of our own life, has thus its origin in barbarian antiquity.

The mountaineers of Caucasia offer another extremely instructive field for illustrations of the same kind. In studying the present customs of the Ossetes—their joint families and communes and their judiciary conceptions—Professor Kovalevsky, in a remarkable work on *Modern Custom and Ancient Law* was enabled step by step to trace the similar dispositions of the old barbarian codes and even to study the origins of feudalism. With other Caucasian stems we occasionally catch a glimpse into the origin of the village community in those cases where it was not tribal but originated from a voluntary union between families of distinct origin. Such was recently the case with some Khevsoure villages, the inhabitants of which took the oath of "community and fraternity."[1] In another part of Caucasus, Daghestan, we see the growth of feudal relations between two tribes, both maintaining at the same time their village communities (and even traces of the gentile "classes"), and thus giving a living illustration of the forms taken by the conquest of Italy and Gaul by the barbarians. The victorious race, the Lezghines, who have conquered several

[1] N. Khoudadoff, "Notes on the Khevsoures," in *Zapiski* of the Caucasian Geogr. Society, xiv. 1, Tiflis, 1890, p. 68. They also took the oath of not marrying girls from their own union, thus displaying a remarkable return to the old gentile rules.

Georgian and Tartar villages in the Zakataly district, did not bring them under the dominion of separate families; they constituted a feudal clan which now includes 12,000 households in three villages, and owns in common no less than twenty Georgian and Tartar villages. The conquerors divided their own land among their clans, and the clans divided it in equal parts among the families; but they did not interfere with the *djemmâas* of their tributaries which still practise the habit mentioned by Julius Cæsar; namely, the *djemmâa* decides each year which part of the communal territory must be cultivated, and this land is divided into as many parts as there are families, and the parts are distributed by lot. It is worthy of note that although proletarians are of common occurrence among the Lezghines (who live under a system of private property in land, and common ownership of serfs [1]) they are rare among their Georgian serfs, who continue to hold their land in common. As to the customary law of the Caucasian mountaineers, it is much the same as that of the Longobards or Salic Franks, and several of its dispositions explain a good deal the judicial procedure of the barbarians of old. Being of a very impressionable character, they do their best to prevent quarrels from taking a fatal issue; so, with the Khevsoures, the swords are very soon drawn when a quarrel breaks out; but if a woman rushes out and throws among them the piece of linen which she wears on her head, the swords are at once returned to their sheaths, and the quarrel is appeased. The head-dress

[1] Dm. Bakradze, "Notes on the Zakataly District," in same *Zapiski*, xiv. 1, p. 264. The "joint team" is as common among the Lezghines as it is among the Ossetes.

of the women is *anaya.* If a quarrel has not been stopped in time and has ended in murder, the compensation money is so considerable that the aggressor is entirely ruined for his life, unless he is adopted by the wronged family; and if he has resorted to his sword in a trifling quarrel and has inflicted wounds, he loses for ever the consideration of his kin. In all disputes, mediators take the matter in hand; they select from among the members of the clan the judges—six in smaller affairs, and from ten to fifteen in more serious matters—and Russian observers testify to the absolute incorruptibility of the judges. An oath has such a significance that men enjoying general esteem are dispensed from taking it: a simple affirmation is quite sufficient, the more so as in grave affairs the Khevsoure never hesitates to recognize his guilt (I mean, of course, the Khevsoure untouched yet by civilization). The oath is chiefly reserved for such cases, like disputes about property, which require some sort of appreciation in addition to a simple statement of facts; and in such cases the men whose affirmation will decide in the dispute, act with the greatest circumspection. Altogether it is certainly not a want of honesty or of respect to the rights of the congeners which characterizes the barbarian societies of Caucasus.

The stems of Africa offer such an immense variety of extremely interesting societies standing at all intermediate stages from the early village community to the despotic barbarian monarchies that I must abandon the idea of giving here even the chief results of a comparative study of their institutions.[1] Suffice it to

[1] See Post, *Afrikanische Jurisprudenz*, Oldenburg, 1887; Münzinger, *Ueber das Recht und Sitten der Bogos*, Winterthur, 1859; Casalis, *Les Bassoutos*, Paris, 1859; Maclean, *Kafir Laws and Customs*, Mount Coke, 1858, etc.

say, that, even under the most horrid despotism of kings, the folkmotes of the village communities and their customary law remain sovereign in a wide circle of affairs. The law of the State allows the king to take any one's life for a simple caprice, or even for simply satisfying his gluttony; but the customary law of the people continues to maintain the same network of institutions for mutual support which exist among other barbarians or have existed among our ancestors. And with some better-favoured stems (in Bornu, Uganda, Abyssinia), and especially the Bogos, some of the dispositions of the customary law are inspired with really graceful and delicate feelings.

The village communities of the natives of both Americas have the same character. The Tupi of Brazil were found living in "long houses" occupied by whole clans which used to cultivate their corn and manioc fields in common. The Arani, much more advanced in civilization, used to cultivate their fields in common; so also the Oucagas, who had learned under their system of primitive communism and "long houses" to build good roads and to carry on a variety of domestic industries,[1] not inferior to those of the early mediæval times in Europe. All of them were also living under the same customary law of which we have given specimens on the preceding pages. At another extremity of the world we find the Malayan feudalism, but this feudalism has been powerless to unroot the *negaria*, or village community, with its common ownership of at least part of the land, and the redistribution of land among the several *negarias* of the tribe.[2] With the Alfurus of Minahasa we find the

[1] Waitz, iii. 423 *seq.*

[2] Post's *Studien zur Entwicklungsgeschichte des Familien-Rechts* Oldenburg, 1889, pp. 270 *seq.*

communal rotation of the crops; with the Indian stem of the Wyandots we have the periodical redistribution of land within the tribe, and the clan-culture of the soil; and in all those parts of Sumatra where Moslem institutions have not yet totally destroyed the old organization we find the joint family (*suka*) and the village community (*kota*) which maintains its right upon the land, even if part of it has been cleared without its authorization.[1] But to say this, is to say that all customs for mutual protection and prevention of feuds and wars, which have been briefly indicated in the preceding pages as characteristic of the village community, exist as well. More than that: the more fully the communal possession of land has been maintained, the better and the gentler are the habits. De Stuers positively affirms that wherever the institution of the village community has been less encroached upon by the conquerors, the inequalities of fortunes are smaller, and the very prescriptions of the *lex talionis* are less cruel; while, on the contrary, wherever the village community has been totally broken up, "the inhabitants suffer the most unbearable oppression from their despotic rulers."[2] This is quite natural. And when Waitz made the remark that those stems which have maintained their tribal confederations stand on a higher level of development and have a richer literature than those stems which have forfeited the old bonds of union, he only pointed out what might have been foretold in advance.

More illustrations would simply involve me in tedious repetitions—so strikingly similar are the barbarian

[1] Powell, *Annual Report of the Bureau of Ethnography*, Washington, 1881, quoted in Post's *Studien*, p. 290; Bastian's *Inselgruppen in Oceanien*, 1883, p. 88.

[2] De Stuers, quoted by Waitz, v. 141.

societies under all climates and amidst all races. The same process of evolution has been going on in mankind with a wonderful similarity. When the clan organization, assailed as it was from within by the separate family, and from without by the dismemberment of the migrating clans and the necessity of taking in strangers of different descent—the village community, based upon a territorial conception, came into existence. This new institution, which had naturally grown out of the preceding one—the clan—permitted the barbarians to pass through a most disturbed period of history without being broken into isolated families which would have succumbed in the struggle for life. New forms of culture developed under the new organization; agriculture attained the stage which it hardly has surpassed until now with the great number; the domestic industries reached a high degree of perfection. The wilderness was conquered, it was intersected by roads, dotted with swarms thrown off by the mother-communities. Markets and fortified centres, as well as places of public worship, were erected. The conceptions of a wider union, extended to whole stems and to several stems of various origin, were slowly elaborated. The old conceptions of justice which were conceptions of mere revenge, slowly underwent a deep modification—the idea of amends for the wrong done taking the place of revenge. The customary law which still makes the law of the daily life for two-thirds or more of mankind, was elaborated under that organization, as well as a system of habits intended to prevent the oppression of the masses by the minorities whose powers grew in proportion to the growing facilities for private accumulation of wealth. This was the new form taken by the tendencies of the masses for mutual support. And the progress—economical, intellectual, and moral—which mankind accomplished under this new popular form of organization, was so great that the States, when they were called later on into existence, simply took possession, in the interest of the minorities, of all the judicial, economical, and administrative functions which the village community already had exercised in the interest of all.

CHAPTER V

MUTUAL AID IN THE MEDIÆVAL CITY

Growth of authority in Barbarian Society.—Serfdom in the villages.—Revolt of fortified towns: their liberation; their charts.—The guild.—Double origin of the free mediæval city.—Self-jurisdiction, self-administration.—Honourable position of labour.—Trade by the guild and by the city.

Sociability and need of mutual aid and support are such inherent parts of human nature that at no time of history can we discover men living in small isolated families, fighting each other for the means of subsistence. On the contrary, modern research, as we saw it in the two preceding chapters, proves that since the very beginning of their prehistoric life men used to agglomerate into *gentes*, clans, or tribes, maintained by an idea of common descent and by worship of common ancestors. For thousands and thousands of years this organization has kept men together, even though there was no authority whatever to impose it. It has deeply impressed all subsequent development of mankind; and when the bonds of common descent had been loosened by migrations on a grand scale, while the development of the separated family within the clan itself had destroyed the old unity of the clan, a new form of union, territorial in its principle—the village community—was called into existence by the social genius of man. This institution, again, kept men together for a number of centuries, permitting them to further develop their social institutions and to pass through some of the darkest periods of history, without being dissolved into loose aggregations of families and individuals, to make a further step in their evolution, and to work out a number of secondary social institutions, several of which have survived down to the present time. We have now to follow the further developments of the same ever-living tendency for mutual aid. Taking the village communities of the so-called barbarians at a time when they were making a new start of civilization after the fall of the Roman Empire, we have to study the new aspects taken by the sociable wants of the masses in the middle ages, and especially in the mediæval guilds and the mediæval city.

Far from being the fighting animals they have often been compared to, the barbarians of the first centuries of our era (like so many Mongolians, Africans, Arabs, and so on, who still continue in the same barbarian stage) invariably preferred peace to war. With the exception of a few tribes which had been driven during the great migrations into unproductive deserts or highlands, and were thus compelled periodically to prey upon their better-favoured neighbours—apart from these, the great bulk of the Teutons, the Saxons, the Celts, the Slavonians, and so on, very soon after they had settled in their newly-conquered abodes, reverted to the spade or to their herds. The earliest barbarian codes already represent to us societies composed of peaceful agricultural communities, not hordes of men at war with each other. These barbarians covered the country with villages and farmhouses;[1] they cleared the forests, bridged the torrents, and colonized the formerly quite uninhabited wilderness; and they left the uncertain warlike pursuits to brotherhoods, *scholæ*, or "trusts" of unruly men, gathered round temporary chieftains, who wandered about, offering their adventurous spirit, their arms, and their knowledge of warfare for the protection of populations, only too anxious to be left in peace. The warrior bands came and went, prosecuting their family feuds; but the great mass continued to till the soil, taking but little notice of their would-be rulers, so long as they did not interfere with the independence of their village communities.[2] The new occupiers of Europe evolved the systems of land tenure and soil culture which are still in force with hundreds of millions of men; they worked out their systems of compensation for wrongs, instead of the old tribal blood-revenge; they learned the first rudiments of industry; and while they fortified their villages with palisaded walls, or erected towers and earthen forts whereto to repair in case of a new invasion, they soon abandoned the task of defending these towers and forts to those who made of war a speciality.

The very peacefulness of the barbarians, certainly not their supposed warlike instincts, thus became the source of their subsequent subjection to the military chieftains. It is evident that the very mode of life of

[1] W. Arnold, in his *Wanderungen und Ansiedelungen der deutschen Stämme*, p. 431, even maintains that one-half of the now arable area in middle Germany must have been reclaimed from the sixth to the ninth century. Nitzsch (*Geschichte des deutschen Volkes*, Leipzig, 1883, vol. i.) shares the same opinion.

[2] Leo and Botta, *Histoire d'Italie*, French edition, 1844, t. i., p. 37.

the armed brotherhoods offered them more facilities for enrichment than the tillers of the soil could find in their agricultural communities. Even now we see that armed men occasionally come together to shoot down Matabeles and to rob them of their droves of cattle, though the Matabeles only want peace and are ready to buy it at a high price. The *scholæ* of old certainly were not more scrupulous than the *scholæ* of our own time. Droves of cattle, iron (which was extremely costly at that time[1]), and slaves were appropriated in this way; and although most acquisitions were wasted on the spot in those glorious feasts of which epic poetry has so much to say—still some part of the robbed riches was used for further enrichment. There was plenty of waste land, and no lack of men ready to till it, if only they could obtain the necessary cattle and implements. Whole villages, ruined by murrains, pests, fires, or raids of new immigrants, were often abandoned by their inhabitants, who went anywhere in search of new abodes. They still do so in Russia in similar circumstances. And if one of the *hirdmen* of the armed brotherhoods offered the peasants some cattle for a fresh start, some iron to make a plough, if not the plough itself, his protection from further raids, and a number of years free from all obligations, before they should begin to repay the contracted debt, they settled upon the land. And when, after a hard fight with bad crops, inundations

[1] The composition for the stealing of a simple knife was 15 *solidi*, and of the iron parts of a mill, 45 *solidi*. (See on this subject Lamprecht's *Wirthschaft und Recht der Franken* in Raumer's *Historisches Taschenbuch*, 1883, p. 52.) According to the Riparian law, the sword, the spear, and the iron armour of a warrior attained the value of at least twenty-five cows, or two years of a freeman's labour. A cuirass alone was valued in the Salic law (Desmichels, quoted by Michelet) at as much as thirty-six bushels of wheat.

and pestilences, those pioneers began to repay their debts, they fell into servile obligations towards the protector of the territory. Wealth undoubtedly did accumulate in this way, and power always follows wealth.[1] And yet, the more we penetrate into the life of those times, the sixth and seventh centuries of our era, the more we see that another element, besides wealth and military force, was required to constitute the authority of the few. It was an element of law and right, a desire of the masses to maintain peace, and to establish what they considered to be justice, which gave to the chieftains of the *scholæ*—kings, dukes, *knyazes*, and the like—the force they acquired two or three hundred years later. That same idea of justice, conceived as an adequate revenge for the wrong done, which had grown in the tribal stage, now passed as a red thread through the history of subsequent institutions, and, much more even than military or economic causes, it became the basis upon which the authority of the kings and the feudal lords was founded.

In fact, one of the chief preoccupations of the barbarian village community always was, as it still is with our barbarian contemporaries, to put a speedy end to the feuds which arose from the then current conception of justice. When a quarrel took place, the community at once interfered, and after the folkmote

[1] The chief wealth of the chieftains, for a long time, was in their personal domains peopled partly with prisoner slaves, but mostly in the above way. On the origin of property see Inama Sternegg's *Die Ausbildung der grossen Grundherrschaften in Deutschland*, in Schmoller's *Forschungen*, Bd. I., 1878; F. Dahn's *Urgeschichte der germanischen und romanischen Völker*, Berlin, 1881; Maurer's *Dorfverfassung;* Guizot's *Essais sur l'histoire de France;* Maine's *Village Community;* Botta's *Histoire d'Italie;* F. Seebohm, Vinogradov, J. R. Green, etc.

had heard the case, it settled the amount of composition (*wergeld*) to be paid to the wronged person, or to his family, as well as the *fred*, or fine for breach of peace, which had to be paid to the community. Interior quarrels were easily appeased in this way. But when feuds broke out between two different tribes, or two confederations of tribes, notwithstanding all measures taken to prevent them,[1] the difficulty was to find an arbiter or sentence-finder whose decision should be accepted by both parties alike, both for his impartiality and for his knowledge of the oldest law. The difficulty was the greater as the customary laws of different tribes and confederations were at variance as to the compensation due in different cases. It therefore became habitual to take the sentence-finder from among such families, or such tribes, as were reputed for keeping the law of old in its purity; of being versed in the songs, triads, sagas, etc., by means of which law was perpetuated in memory; and to retain law in this way became a sort of art, a "mystery," carefully transmitted in certain families from generation to generation. Thus in Iceland, and in other Scandinavian lands, at every *Allthing*, or national folkmote, a *lövsögmathr* used to recite the whole law from memory for the enlightening of the assembly; and in Ireland there was, as is known, a special class of men reputed for the knowledge of the old traditions, and therefore enjoying a great authority as judges.[2] Again, when we are told by the Russian annals that some stems of North-West Russia, moved by the

[1] See Sir Henry Maine's *International Law*, London, 1888.

[2] *Ancient Laws of Ireland*, Introduction; E. Nys, *Études de droit international*, t. i., 1896, pp. 86 *seq.* Among the Ossetes the arbiters from three *oldest* villages enjoy a special reputation (M. Kovalevsky's *Modern Custom and Old Law*, Moscow, 1886, ii. 217, Russian).

growing disorder which resulted from "clans rising against clans," appealed to Norman *varingiar* to be their judges and commanders of warrior *scholæ;* and when we see the *knyazes*, or dukes, elected for the next two hundred years always from the same Norman family, we cannot but recognize that the Slavonians trusted to the Normans for a better knowledge of the law which would be equally recognized as good by different Slavonian kins. In this case the possession of runes, used for the transmission of old customs, was a decided advantage in favour of the Normans; but in other cases there are faint indications that the "eldest" branch of the stem, the supposed mother-branch, was appealed to to supply the judges, and its decisions were relied upon as just;[1] while at a later epoch we see a distinct tendency towards taking the sentence-finders from the Christian clergy, which, at that time, kept still to the fundamental, now forgotten, principle of Christianity, that retaliation is no act of justice. At that time the Christian clergy opened the churches as places of asylum for those who fled from blood revenge, and they willingly acted as arbiters in criminal cases, always opposing the old tribal principle of life for life and wound for wound. In short, the deeper we penetrate into the history of early institutions, the less we find grounds for the military theory of origin of authority. Even that power which later on became such a source of oppression seems, on the contrary, to have found its origin in the peaceful inclinations of the masses.

In all these cases the *fred*, which often amounted

[1] It is permissible to think that this conception (related to the conception of tanistry) played an important part in the life of the period; but research has not yet been directed that way.

to half the compensation, went to the folkmote, and from times immemorial it used to be applied to works of common utility and defence. It has still the same destination (the erection of towers) among the Kabyles and certain Mongolian stems; and we have direct evidence that even several centuries later the judicial fines, in Pskov and several French and German cities, continued to be used for the repair of the city walls.[1] It was thus quite natural that the fines should be handed over to the sentence-finder, who was bound, in return, both to maintain the *schola* of armed men to whom the defence of the territory was trusted, and to execute the sentences. This became a universal custom in the eighth and ninth centuries, even when the sentence-finder was an elected bishop. The germ of a combination of what we should now call the judicial power and the executive thus made its appearance. But to these two functions the attributions of the duke or king were strictly limited. He was no ruler of the people—the supreme power still belonging to the folkmote—not even a commander of the popular militia; when the folk took to arms, it marched under a separate, also elected, commander, who was not a subordinate, but an equal to the king.[2] The king was a lord on his personal domain only. In fact, in barbarian language, the word *konung*, *koning*, or *cyning*, synonymous with the Latin *rex*, had no other meaning than that of a temporary leader or chieftain

[1] It was distinctly stated in the charter of St. Quentin of the year 1002 that the ransom for houses which had to be demolished for crimes went for the city walls. The same destination was given to the *Ungeld* in German cities. At Pskov the cathedral was the bank for the fines, and from this fund money was taken for the walls.

[2] Sohm, *Fränkische Rechts- und Gerichtsverfassung*, p. 23; also Nitzsch, *Geschichte des deutschen Volkes*, i. 78.

of a band of men. The commander of a flotilla of boats, or even of a single pirate boat, was also a *konung*, and till the present day the commander of fishing in Norway is named *Not-kong*—"the king of the nets."[1] The veneration attached later on to the personality of a king did not yet exist, and while treason to the kin was punished by death, the slaying of a king could be recouped by the payment of compensation: a king simply was valued so much more than a freeman.[2] And when King Knu (or Canute) had killed one man of his own *schola*, the saga represents him convoking his comrades to a *thing* where he knelt down imploring pardon. He was pardoned, but not till he had agreed to pay nine times the regular composition, of which one-third went to himself for the loss of one of his men, one-third to the relatives of the slain man, and one-third (the *fred*) to the *schola*.[3] In reality, a complete change had to be accomplished in the current conceptions, under the double influence of the Church and the students of Roman law, before an idea of sanctity began to be attached to the personality of the king.

However, it lies beyond the scope of these essays

[1] See the excellent remarks on this subject in Augustin Thierry's *Lettres sur l'histoire de France*, 7th letter. The barbarian translations of parts of the Bible are extremely instructive on this point.

[2] Thirty-six times more than a noble, according to the Anglo-Saxon law. In the code of Rothari the slaying of a king is, however, punished by death; but (apart from Roman influence) this new disposition was introduced (in 646) in the Lombardian law—as remarked by Leo and Botta—to cover the king from blood revenge. The king being at that time the executioner of his own sentences (as the tribe formerly was of its own sentences), he had to be protected by a special disposition, the more so as several Lombardian kings before Rothari had been slain in succession (Leo and Botta, *l. c.*, i. 66–90).

[3] Kaufmann, *Deutsche Geschichte*, Bd. I. "Die Germanen der Urzeit," p. 133.

to follow the gradual development of authority out of the elements just indicated. Historians, such as Mr. and Mrs. Green for this country, Augustin Thierry, Michelet, and Luchaire for France, Kaufmann, Janssen, W. Arnold, and even Nitzsch, for Germany, Leo and Botta for Italy, Byelaeff, Kostomaroff, and their followers for Russia, and many others, have fully told that tale. They have shown how populations, once free, and simply agreeing "to feed" a certain portion of their military defenders, gradually became the serfs of these protectors; how "commendation" to the Church, or to a lord, became a hard necessity for the freeman; how each lord's and bishop's castle became a robber's nest—how feudalism was imposed, in a word—and how the crusades, by freeing the serfs who wore the cross, gave the first impulse to popular emancipation. All this need not be retold in this place, our chief aim being to follow the *constructive* genius of the masses in their mutual-aid institutions.

At a time when the last vestiges of barbarian freedom seemed to disappear, and Europe, fallen under the dominion of thousands of petty rulers, was marching towards the constitution of such theocracies and despotic States as had followed the barbarian stage during the previous starts of civilization, or of barbarian monarchies, such as we see now in Africa, life in Europe took another direction. It went on on lines similar to those it had once taken in the cities of antique Greece. With a unanimity which seems almost incomprehensible, and for a long time was not understood by historians, the urban agglomerations, down to the smallest burgs, began to shake off the yoke of their worldly and clerical lords. The fortified village rose against the lord's castle, defied it first, attacked it next, and finally destroyed it. The movement spread from spot to spot, involving every town on the surface of Europe, and in less than a hundred years free cities had been called into existence on the coasts of the Mediterranean, the North Sea, the Baltic, the Atlantic Ocean, down to the fjords of Scandinavia; at the feet of the Apennines, the Alps, the Black Forest, the Grampians, and the Carpathians; in the plains of Russia, Hungary, France and Spain. Everywhere the same revolt took place, with the same features, passing through the same phases, leading to the same results. Wherever men had found, or expected to find, some protection behind their town walls, they instituted their "co-jurations," their "fraternities," their "friendships," united in one common idea, and boldly marching towards a new life of mutual support and liberty. And they succeeded so well that in three or four hundred years they had changed the very face of Europe. They had covered the country with beautiful sumptuous buildings, expressing the genius of free unions of free men, unrivalled since for their beauty and expressiveness; and they bequeathed to the following generations all the arts, all the industries, of which our present civilization, with all its achievements and promises for the future, is only a further development. And when we now look to the forces which have produced these grand results, we find them—not in the genius of individual heroes, not in the mighty organization of huge States or the political capacities of their rulers, but in the very same current of mutual aid and support which we saw at work in the village community, and which was vivified and reinforced in the Middle Ages by a new form of unions, inspired by the very same spirit but shaped on a new model—the guilds.

It is well known by this time that feudalism did not imply a dissolution of the village community. Although the lord had succeeded in imposing servile labour upon the peasants, and had appropriated for himself such rights as were formerly vested in the village community alone (taxes, mortmain, duties on inheritances and marriages), the peasants had, nevertheless, maintained the two fundamental rights of their communities: the common possession of the land, and self-jurisdiction. In olden times, when a king sent his vogt to a village, the peasants received him with flowers in one hand and arms in the other, and asked him—which law he intended to apply: the one he found in the village, or the one he brought with him? And, in the first case, they handed him the flowers and accepted him; while in the second case they fought him.[1] Now, they accepted the king's or the lord's official whom they could not refuse; but they maintained the folkmote's jurisdiction, and themselves nominated six, seven, or twelve judges, who acted with the lord's judge, in the presence of the folkmote, as arbiters and sentence-finders. In most cases the official had nothing left to him but to confirm the sentence and to levy the customary *fred.* This precious right of self-jurisdiction, which, at that time, meant self-administration and self-legislation, had been maintained through all the struggles; and even the lawyers by whom Karl the Great was surrounded could not abolish it; they were bound to confirm it. At the same time, in all matters concerning the com-

[1] Dr. F. Dahn, *Urgeschichte der germanischen und romanischen Völker*, Berlin, 1881, Bd. I. 96.

munity's domain, the folkmote retained its supremacy and (as shown by Maurer) often claimed submission from the lord himself in land tenure matters. No growth of feudalism could break this resistance; the village community kept its ground; and when, in the ninth and tenth centuries, the invasions of the Normans, the Arabs, and the Ugrians had demonstrated that military *scholæ* were of little value for protecting the land, a general movement began all over Europe for fortifying the villages with stone walls and citadels. Thousands of fortified centres were then built by the energies of the village communities; and, once they had built their walls, once a common interest had been created in this new sanctuary—the town walls—they soon understood that they could henceforward resist the encroachments of the inner enemies, the lords, as well as the invasions of foreigners. A new life of freedom began to develop within the fortified enclosures. The mediæval city was born.[1]

[1] If I thus follow the views long since advocated by Maurer (*Geschichte der Städteverfassung in Deutschland*, Erlangen, 1869), it is because he has fully proved the uninterrupted evolution from the village community to the mediæval city, and that his views alone can explain the universality of the communal movement. Savigny and Eichhorn and their followers have certainly proved that the traditions of the Roman *municipia* had never totally disappeared. But they took no account of the village-community period which the barbarians lived through before they had any cities. The fact is, that whenever mankind made a new start in civilization, in Greece, Rome, or middle Europe, it passed through the same stages—the tribe, the village community, the free city, the state—each one naturally evolving out of the preceding stage. Of course, the experience of each preceding civilization was never lost. Greece (itself influenced by Eastern civilizations) influenced Rome, and Rome influenced our civilization; but each of them began from the same beginning—the tribe. And just as we cannot say that our states are *continuations* of the Roman state, so also can we not say that the mediæval cities of Europe (including Scandinavia and

No period of history could better illustrate the constructive powers of the popular masses than the tenth and eleventh centuries, when the fortified villages and market-places, representing so many "oases amidst the feudal forest," began to free themselves from their lord's yoke, and slowly elaborated the future city organization; but, unhappily, this is a period about which historical information is especially scarce: we know the results, but little has reached us about the means by which they were achieved. Under the protection of their walls the cities' folkmotes—either quite independent, or led by the chief noble or merchant families—conquered and maintained the right of electing the military *defensor* and supreme judge of the town, or at least of choosing between those who pretended to occupy this position. In Italy the young communes were continually sending away their *defensors* or *domini*, fighting those who refused to go. The same went on in the East. In Bohemia, rich and poor alike (*Bohemicæ gentis magni et parvi, nobiles et ignobiles*) took part in the election;[1] while the *vyeches* (folkmotes) of the Russian cities regularly elected their dukes—always from the same Rurik family—covenanted with them, and sent the *knyaz* away if he had provoked discontent.[2] At the same

---

Russia) were a continuation of the Roman cities. They were a continuation of the barbarian village community, influenced to a certain extent by the traditions of the Roman towns.

[1] M. Kovalevsky, *Modern Customs and Ancient Laws of Russia* (Ilchester Lectures, London, 1891, lecture 4).

[2] A considerable amount of research had to be done before this character of the so-called *udyelnyi period* was properly established by the works of Byelaeff (*Tales from Russian History*), Kostomaroff (*The Beginnings of Autocracy in Russia*), and especially Professor Sergievich (*The Vyeche and the Prince*). The English reader may find some information about this period in the just-named work of

time in most cities of Western and Southern Europe, the tendency was to take for *defensor* a bishop whom the city had elected itself; and so many bishops took the lead in protecting the "immunities" of the towns and in defending their liberties, that numbers of them were considered, after their death, as saints and special patrons of different cities. St. Uthelred of Winchester, St. Ulrik of Augsburg, St. Wolfgang of Ratisbon, St. Heribert of Cologne, St. Adalbert of Prague, and so on, as well as many abbots and monks, became so many cities' saints for having acted in defence of popular rights.[1] And under the new *defensors*, whether laic or clerical, the citizens conquered full self-jurisdiction and self-administration for their folkmotes.[2]

The whole process of liberation progressed by a series of imperceptible acts of devotion to the common cause, accomplished by men who came out of the masses—by unknown heroes whose very names have not been preserved by history. The wonderful movement of the God's peace (*treuga Dei*) by which the popular masses endeavoured to put a limit to the endless family feuds of the noble families, was born in the young towns, the bishops and the citizens trying to extend to the nobles the peace they had established

M. Kovalevsky, in Rambaud's *History of Russia*, and, in a short summary, in the article "Russia" of the last edition of *Chambers's Encyclopædia*.

[1] Ferrari, *Histoire des révolutions d'Italie*, i. 257; Kallsen, *Die deutschen Städte im Mittelalter*, Bd. I. (Halle, 1891).

[2] See the excellent remarks of Mr. G. L. Gomme as regards the folkmote of London (*The Literature of Local Institutions*, London, 1886, p. 76). It must, however, be remarked that in royal cities the folkmote never attained the independence which it assumed elsewhere. It is even certain that Moscow and Paris were chosen by the kings and the Church as the cradles of the future royal authority in the State, because they did not possess the tradition of folkmotes accustomed to act as sovereign in all matters.

within their town walls.[1] Already at that period, the commercial cities of Italy, and especially Amalfi (which had its elected consuls since 844, and frequently changed its doges in the tenth century)[2] worked out the customary maritime and commercial law which later on became a model for all Europe; Ravenna elaborated its craft organization, and Milan, which had made its first revolution in 980, became a great centre of commerce, its trades enjoying a full independence since the eleventh century.[3] So also Bruges and Ghent; so also several cities of France in which the *Mahl* or *forum* had become a quite independent institution.[4] And already during that period began the work of artistic decoration of the towns by works of architecture, which we still admire and which loudly testify of the intellectual movement of the times. "The basilicæ were then renewed in almost all the universe," Raoul Glaber wrote in his chronicle, and some of the finest monuments of mediæval architecture date from that period: the wonderful old church of Bremen was built in the ninth century, Saint Marc of Venice was finished in 1071, and the beautiful dome of Pisa in 1063. In fact, the intel-

[1] A. Luchaire, *Les Communes françaises;* also Kluckohn, *Geschichte des Gottesfrieden*, 1857. L. Sémichon (*La paix et la trève de Dieu*, 2 vols., Paris, 1869) has tried to represent the communal movement as issued from that institution. In reality, the *treuga Dei*, like the league started under Louis le Gros for the defence against both the robberies of the nobles and the Norman invasions, was a thoroughly *popular* movement. The only historian who mentions this last league—that is, Vitalis—describes it as a "popular community" ("Considérations sur l'histoire de France," in vol. iv. of Aug. Thierry's *Œuvres*, Paris, 1868, p. 191 and *note*).

[2] Ferrari, i. 152, 263, etc.

[3] Perrens, *Histoire de Florence*, i. 188; Ferrari, *l. c.*, i. 283.

[4] Aug. Thierry, *Essai sur l'histoire du Tiers État*, Paris, 1875, p. 414, *note*.

lectual movement which has been described as the Twelfth Century Renaissance[1] and the Twelfth Century Rationalism—the precursor of the Reform[2]—date from that period, when most cities were still simple agglomerations of small village communities enclosed by walls.

However, another element, besides the village-community principle, was required to give to these growing centres of liberty and enlightenment the unity of thought and action, and the powers of initiative, which made their force in the twelfth and thirteenth centuries. With the growing diversity of occupations, crafts and arts, and with the growing commerce in distant lands, some new form of union was required, and this necessary new element was supplied by the *guilds*. Volumes and volumes have been written about these unions which, under the name of guilds, brotherhoods, friendships and *druzhestva*, *minne*, *artels* in Russia, *esnaifs* in Servia and Turkey, *amkari* in Georgia, and so on, took such a formidable development in mediæval times and played such an important part in the emancipation of the cities. But it took historians more than sixty years before the universality of this institution and its true characters were understood. Only now, when hundreds of guild statutes have been published and studied, and their relationship to the Roman *collegia*, and the earlier unions in Greece and in India,[3] is known, can

[1] F. Rocquain, "La Renaissance au XIIe siècle," in *Études sur l'histoire de France*, Paris, 1875, pp. 55–117.

[2] N. Kostomaroff, "The Rationalists of the Twelfth Century," in his *Monographies and Researches* (Russian).

[3] Very interesting facts relative to the universality of guilds will be found in "Two Thousand Years of Guild Life," by Rev. J. M.

we maintain with full confidence that these brotherhoods were but a further development of the same principles which we saw at work in the *gens* and the village community.

Nothing illustrates better these mediæval brotherhoods than those temporary guilds which were formed on board ships. When a ship of the Hansa had accomplished her first half-day passage after having left the port, the captain (*Schiffer*) gathered all crew and passengers on the deck, and held the following language, as reported by a contemporary :—

"'As we are now at the mercy of God and the waves,' he said, 'each one must be equal to each other. And as we are surrounded by storms, high waves, pirates and other dangers, we must keep a strict order that we may bring our voyage to a good end. That is why we shall pronounce the prayer for a good wind and good success, and, according to marine law, we shall name the occupiers of the judges' seats (*Schöffenstellen*).' Thereupon the crew elected a Vogt and four *scabini*, to act as their judges. At the end of the voyage the Vogt and the *scabini* abdicated their functions and addressed the crew as follows :—'What has happened on board ship, we must pardon to each other and consider as dead (*todt und ab sein lassen*). What we have judged right, was for the sake of justice. This is why we beg you all, in the name of honest justice, to forget all the animosity one may nourish against another, and to swear on bread and salt that he will not think of it in a bad spirit. If any one, however, considers himself wronged, he must appeal to the land Vogt and ask justice from him before sunset.' On landing, the Stock with the *fred*-fines was handed over to the Vogt of the sea-port for distribution among the poor."[1]

This simple narrative, perhaps better than anything

Lambert, Hull, 1891. On the Georgian *amkari*, see S. Eghiazarov, *Gorodskiye Tsekhi* ("Organization of Transcaucasian Amkari"), in *Memoirs* of the Caucasian Geographical Society, xiv. 2, 1891.

[1] J. D. Wunderer's "Reisebericht" in Fichard's *Frankfurter Archiv*, ii. 245 ; quoted by Janssen, *Geschichte des deutschen Volkes*, i. 355.

else, depicts the spirit of the mediæval guilds. Like organizations came into existence wherever a group of men—fishermen, hunters, travelling merchants, builders, or settled craftsmen—came together for a common pursuit. Thus, there was on board ship the naval authority of the captain; but, for the very success of the common enterprise, all men on board, rich and poor, masters and crew, captain and sailors, agreed to be equals in their mutual relations, to be simply men, bound to aid each other and to settle their possible disputes before judges elected by all of them. So also when a number of craftsmen—masons, carpenters, stone-cutters, etc.—came together for building, say, a cathedral, they all belonged to a city which had its political organization, and each of them belonged moreover to his own craft; but they were united besides by their common enterprise, which they knew better than any one else, and they joined into a body united by closer, although temporary, bonds; they founded the guild for the building of the cathedral.[1] We may see the same till now in the Kabylian *çof*:[2] the Kabyles have their village community; but this union is not sufficient for all political, commercial, and personal needs of union, and the closer brotherhood of the *çof* is constituted.

As to the social characters of the mediæval guild, any guild-statute may illustrate them. Taking, for instance, the *skraa* of some early Danish guild, we read in it, first, a statement of the general brotherly feelings which must reign in the guild; next come the regulations relative to self-jurisdiction in cases of quarrels

[1] Dr. Leonard Ennen, *Der Dom zu Köln, Historische Einleitung*, Köln, 1871, pp. 46, 50.
[2] See previous chapter.

arising between two brothers, or a brother and a stranger; and then, the social duties of the brethren are enumerated. If a brother's house is burned, or he has lost his ship, or has suffered on a pilgrim's voyage, all the brethren must come to his aid. If a brother falls dangerously ill, two brethren must keep watch by his bed till he is out of danger, and if he dies, the brethren must bury him—a great affair in those times of pestilences—and follow him to the church and the grave. After his death they must provide for his children, if necessary; very often the widow becomes a sister to the guild.[1]

These two leading features appeared in every brotherhood formed for any possible purpose. In each case the members treated each other as, and named each other, brother and sister;[2] all were equals before the guild. They owned some "chattel" (cattle, land, buildings, places of worship, or "stock") in common. All brothers took the oath of abandoning all feuds of old; and, without imposing upon each other the obligation of never quarrelling again, they agreed that no quarrel should degenerate into a feud, or into a law-suit before another court than the tribunal of the brothers themselves. And if a brother was involved in a quarrel with a stranger to the guild, they agreed to support him for bad and for good; that is, whether he was unjustly accused of aggression, or really was the aggressor, they had to support him, and to bring

[1] Kofod Ancher, *Om gamle Danske Gilder og deres Undergång*, Copenhagen, 1785. Statutes of a Knu guild.

[2] Upon the position of women in guilds, see Miss Toulmin Smith's introductory remarks to the *English Guilds* of her father. One of the Cambridge statutes (p. 281) of the year 1503 is quite positive in the following sentence: "Thys statute is made by the comyne assent of all the bretherne and sisterne of alhallowe yelde."

things to a peaceful end. So long as his was not a secret aggression—in which case he would have been treated as an outlaw—the brotherhood stood by him.[1] If the relatives of the wronged man wanted to revenge the offence at once by a new aggression, the brotherhood supplied him with a horse to run away, or with a boat, a pair of oars, a knife and a steel for striking light; if he remained in town, twelve brothers accompanied him to protect him; and in the meantime they arranged the composition. They went to court to support by oath the truthfulness of his statements, and if he was found guilty they did not let him go to full ruin and become a slave through not paying the due compensation: they all paid it, just as the *gens* did in olden times. Only when a brother had broken the faith towards his guild-brethren, or other people, he was excluded from the brotherhood "with a Nothing's name" (*tha scal han maeles af brödrescap met nidings nafn*).[2]

Such were the leading ideas of those brotherhoods which gradually covered the whole of mediæval life. In fact, we know of guilds among all possible professions: guilds of serfs,[3] guilds of freemen, and guilds of both serfs and freemen; guilds called into life for the special purpose of hunting, fishing, or a trading

[1] In mediæval times, only secret aggression was treated as a murder. Blood-revenge in broad daylight was justice; and slaying in a quarrel was not murder, once the aggressor showed his willingness to repent and to repair the wrong he had done. Deep traces of this distinction still exist in modern criminal law, especially in Russia.

[2] Kofod Ancher, *l. c.* This old booklet contains much that has been lost sight of by later explorers.

[3] They played an important part in the revolts of the serfs, and were therefore prohibited several times in succession in the second half of the ninth century. Of course, the king's prohibitions remained a dead letter.

expedition, and dissolved when the special purpose had been achieved; and guilds lasting for centuries in a given craft or trade. And, in proportion as life took an always greater variety of pursuits, the variety in the guilds grew in proportion. So we see not only merchants, craftsmen, hunters, and peasants united in guilds; we also see guilds of priests, painters, teachers of primary schools and universities, guilds for performing the passion play, for building a church, for developing the "mystery" of a given school of art or craft, or for a special recreation—even guilds among beggars, executioners, and lost women, all organized on the same double principle of self-jurisdiction and mutual support.[1] For Russia we have positive evidence showing that the very "making of Russia" was as much the work of its hunters', fishermen's, and traders' *artels* as of the budding village communities, and up to the present day the country is covered with *artels*.[2]

These few remarks show how incorrect was the view taken by some early explorers of the guilds when they wanted to see the essence of the institution in its

[1] The mediæval Italian painters were also organized in guilds, which became at a later epoch Academies of art. If the Italian art of those times is impressed with so much individuality that we distinguish, even now, between the different schools of Padua, Bassano, Treviso, Verona, and so on, although all these cities were under the sway of Venice, this was due—J. Paul Richter remarks—to the fact that the painters of each city belonged to a separate guild, friendly with the guilds of other towns, but leading a separate existence. The oldest guild-statute known is that of Verona, dating from 1303, but evidently copied from some much older statute. "Fraternal assistance in necessity of whatever kind," "hospitality towards strangers, when passing through the town, as thus information may be obtained about matters which one may like to learn," and "obligation of offering comfort in case of debility" are among the obligations of the members (*Nineteenth Century*, Nov. 1890, and Aug. 1892).

[2] The chief works on the *artels* are named in the article "Russia" of the *Encyclopædia Britannica*, 9th edition, p. 84.

yearly festival. In reality, the day of the common meal was always the day, or the morrow of the day, of election of aldermen, of discussion of alterations in the statutes, and very often the day of judgment of quarrels that had risen among the brethren,[1] or of renewed allegiance to the guild. The common meal, like the festival at the old tribal folkmote—the *mahl* or *malum*—or the Buryate *aba*, or the parish feast and the harvest supper, was simply an affirmation of brotherhood. It symbolized the times when everything was kept in common by the clan. This day, at least, all belonged to all; all sate at the same table and partook of the same meal. Even at a much later time the inmate of the almshouse of a London guild sat this day by the side of the rich alderman. As to the distinction which several explorers have tried to establish between the old Saxon "frith guild" and the so-called "social" or "religious" guilds—all were frith guilds in the sense above mentioned,[2] and all were

[1] See, for instance, the texts of the Cambridge guilds given by Toulmin Smith (*English Guilds*, London, 1870, pp. 274–276), from which it appears that the "generall and principall day" was the "eleccioun day;" or, Ch. M. Clode's *The Early History of the Guild of the Merchant Taylors*, London, 1888, i. 45; and so on. For the renewal of allegiance, see the Jómsviking saga, mentioned in Pappenheim's *Altdänische Schutzgilden*, Breslau, 1885, p. 67. It appears very probable that when the guilds began to be prosecuted, many of them inscribed in their statutes the meal day only, or their pious duties, and only alluded to the judicial function of the guild in vague words; but this function did not disappear till a very much later time. The question, "Who will be my judge?" has no meaning now, since the State has appropriated for its bureaucracy the organization of justice; but it was of primordial importance in mediæval times, the more so as self-jurisdiction meant self-administration. It must also be remarked that the translation of the Saxon and Danish "guild-bretheren," or "brodræ," by the Latin *convivii* must also have contributed to the above confusion.

[2] See the excellent remarks upon the frith guild by J. R. Green and Mrs Green in *The Conquest of England*, London, 1883, pp. 229–230.

religious in the sense in which a village community or a city placed under the protection of a special saint is social and religious. If the institution of the guild has taken such an immense extension in Asia, Africa, and Europe, if it has lived thousands of years, reappearing again and again when similar conditions called it into existence, it is because it was much more than an eating association, or an association for going to church on a certain day, or a burial club. It answered to a deeply inrooted want of human nature; and it embodied all the attributes which the State appropriated later on for its bureaucracy and police, and much more than that. It was an association for mutual support in all circumstances and in all accidents of life, "by deed and advise," and it was an organization for maintaining justice—with this difference from the State, that on all these occasions a humane, a brotherly element was introduced instead of the formal element which is the essential characteristic of State interference. Even when appearing before the guild tribunal, the guild-brother answered before men who knew him well and had stood by him before in their daily work, at the common meal, in the performance of their brotherly duties: men who were his equals and brethren indeed, not theorists of law nor defenders of some one else's interests.[1]

It is evident that an institution so well suited to serve the need of union, without depriving the individual of his initiative, could but spread, grow, and fortify. The difficulty was only to find such form as would permit to federate the unions of the guilds without interfering with the unions of the village com-

[1] See Appendix X.

munities, and to federate all these into one harmonious whole. And when this form of combination had been found, and a series of favourable circumstances permitted the cities to affirm their independence, they did so with a unity of thought which can but excite our admiration, even in our century of railways, telegraphs, and printing. Hundreds of charters in which the cities inscribed their liberation have reached us, and through all of them—notwithstanding the infinite variety of details, which depended upon the more or less greater fulness of emancipation—the same leading ideas run. The city organized itself as a federation of both small village communities and guilds.

"All those who belong to the friendship of the town"—so runs a charter given in 1188 to the burghesses of Aire by Philip, Count of Flanders—"have promised and confirmed by faith and oath that they will aid each other as brethren, in whatever is useful and honest. That if one commits against another an offence in words or in deeds, the one who has suffered therefrom will not take revenge, either himself or his people . . . he will lodge a complaint and the offender will make good for his offence, according to what will be pronounced by twelve elected judges acting as arbiters. And if the offender or the offended, after having been warned thrice, does not submit to the decision of the arbiters, he will be excluded from the friendship as a wicked man and a perjuror.[1] "Each one of the men of the commune will be faithful to his con-juror, and will give him aid and advice, according to what justice will dictate him"—the Amiens and Abbeville charters say. "All will aid each other, according to their powers, within the boundaries of the Commune, and will not suffer that any one takes anything from any one of them, or makes one pay contributions"—do we read in the charters of Soissons, Compiègne, Senlis, and many others of the same type.[2] And so on with countless variations on the same theme.

[1] *Recueil des ordonnances des rois de France*, t. xii. 562; quoted by Aug. Thierry in *Considérations sur l'histoire de France*, p. 196, ed. 12mo.

[2] A. Luchaire, *Les Communes françaises*, pp. 45–46.

"The Commune," Guilbert de Nogent wrote, "is an oath of mutual aid (*mutui adjutorii conjuratio*) . . . A new and detestable word. Through it the serfs (*capite sensi*) are freed from all serfdom; through it, they can only be condemned to a legally determined fine for breaches of the law; through it, they cease to be liable to payments which the serfs always used to pay."[1]

The same wave of emancipation ran, in the twelfth century, through all parts of the continent, involving both rich cities and the poorest towns. And if we may say that, as a rule, the Italian cities were the first to free themselves, we can assign no centre from which the movement would have spread. Very often a small burg in central Europe took the lead for its region, and big agglomerations accepted the little town's charter as a model for their own. Thus, the charter of a small town, Lorris, was adopted by eighty-three towns in south-west France, and that of Beaumont became the model for over five hundred towns and cities in Belgium and France. Special deputies were dispatched by the cities to their neighbours to obtain a copy from their charter, and the constitution was framed upon that model. However, they did not simply copy each other: they framed their own charters in accordance with the concessions they had obtained from their lords; and the result was that, as remarked by an historian, the charters of the mediæval communes offer the same variety as the Gothic architecture of their churches and cathedrals. The same leading ideas in all of them—the cathedral symbolizing the union of parish and guild in the city,—and the same infinitely rich variety of detail.

Self-jurisdiction was the essential point, and self-jurisdiction meant self-administration. But the com-

[1] Guilbert de Nogent, *De vita sua*, quoted by Luchaire, *l. c.*, p. 14.

mune was not simply an "autonomous" part of the State—such ambiguous words had not yet been invented by that time—it was a State in itself. It had the right of war and peace, of federation and alliance with its neighbours. It was sovereign in its own affairs, and mixed with no others. The supreme political power could be vested entirely in a democratic forum, as was the case in Pskov, whose *vyeche* sent and received ambassadors, concluded treaties, accepted and sent away princes, or went on without them for dozens of years; or it was vested in, or usurped by, an aristocracy of merchants or even nobles, as was the case in hundreds of Italian and middle European cities. The principle, nevertheless, remained the same: the city was a State and—what was perhaps still more remarkable—when the power in the city was usurped by an aristocracy of merchants or even nobles, the inner life of the city and the democratism of its daily life did not disappear: they depended but little upon what may be called the political form of the State.

The secret of this seeming anomaly lies in the fact that a mediæval city was not a centralized State. During the first centuries of its existence, the city hardly could be named a State as regards its interior organization, because the middle ages knew no more of the present centralization of functions than of the present territorial centralization. Each group had its share of sovereignty. The city was usually divided into four quarters, or into five to seven sections radiating from a centre, each quarter or section roughly corresponding to a certain trade or profession which prevailed in it, but nevertheless containing inhabitants of different social positions and occupations—nobles, merchants, artisans, or even half-serfs; and each section or quarter constituted a quite independent agglomeration. In Venice, each island was an independent political community. It had its own organized trades, its own commerce in salt, its own jurisdiction and administration, its own forum; and the nomination of a doge by the city changed nothing in the inner independence of the units.[1] In Cologne, we see the inhabitants divided into *Geburschaften* and *Heimschaften* (*viciniæ*), *i. e.* neighbour guilds, which dated from the Franconian period. Each of them had its judge (*Burrichter*) and the usual twelve elected sentence-finders (*Schöffen*), its Vogt, and its *greve* or commander of the local militia.[2] The story of early London before the Conquest—Mr. Green says—is that "of a number of little groups scattered here and there over the area within the walls, each growing up with its own life and institutions, guilds, sokes, religious houses and the like, and only slowly drawing together into a municipal union."[3] And if we refer to the annals of the Russian cities, Novgorod and Pskov, both of which are relatively rich in local details, we find the section (*konets*) consisting of independent streets (*ulitsa*), each of which, though chiefly peopled with artisans of a certain craft, had also merchants and landowners among its inhabitants, and was a separate community. It had the communal responsibility of all members in case of crime, its own jurisdiction and administration by street aldermen (*ulichanskiye starosty*), its own seal and, in case of need, its own

[1] Lebret, *Histoire de Venise*, i. 393; also Marin, quoted by Leo and Botta in *Histoire de l'Italie*, French edition, 1844, t. i. 500.

[2] Dr. W. Arnold, *Verfassungsgeschichte der deutschen Freistädte*, 1854, Bd. ii. 227 *seq.*; Ennen, *Geschichte der Stadt Koeln*, Bd. i. 228–229; also the documents published by Ennen and Eckert.

[3] *Conquest of England*, 1883, p. 453.

forum; its own militia, as also its self-elected priests and its own collective life and collective enterprise.[1]

The mediæval city thus appears as a double federation: of all householders united into small territorial unions—the street, the parish, the section—and of individuals united by oath into guilds according to their professions; the former being a produce of the village-community origin of the city, while the second is a subsequent growth called to life by new conditions.

To guarantee liberty, self-administration, and peace was the chief aim of the mediæval city; and labour, as we shall presently see when speaking of the craft guilds, was its chief foundation. But "production" did not absorb the whole attention of the mediæval economist. With his practical mind, he understood that "consumption" must be guaranteed in order to obtain production; and therefore, to provide for "the common first food and lodging of poor and rich alike" (*gemeine notdurft vnd gemach armer vnd richer*[2]) was the fundamental principle in each city. The purchase of food supplies and other first necessaries (coal, wood, etc.) before they had reached the market, or altogether in especially favourable conditions from which others would be excluded—the *preempcio*, in a word—was entirely prohibited. Everything had to go to the market and be offered there for every one's purchase, till the ringing of the bell had closed the market. Then only could the retailer buy the remainder, and even then his profit should be an "honest

[1] Byelaeff, *Russian History*, vols. ii. and iii.

[2] W. Gramich, *Verfassungs- und Verwaltungsgeschichte der Stadt Würzburg im* 13. *bis zum* 15. *Jahrhundert*, Würzburg, 1882, p. 34.

profit" only.[1] Moreover, when corn was bought by a baker wholesale after the close of the market, every citizen had the right to claim part of the corn (about half-a-quarter) for his own use, at wholesale price, if he did so before the final conclusion of the bargain; and reciprocally, every baker could claim the same if the citizen purchased corn for re-selling it. In the first case, the corn had only to be brought to the town mill to be ground in its proper turn for a settled price, and the bread could be baked in the *four banal*, or communal oven.[2] In short, if a scarcity visited the city, all had to suffer from it more or less; but apart from the calamities, so long as the free cities existed no one could die in their midst from starvation, as is unhappily too often the case in our own times.

However, all such regulations belong to later periods of the cities' life, while at an earlier period it was the city itself which used to buy all food supplies for the use of the citizens. The documents recently published by Mr. Gross are quite positive on this point and fully support his conclusion to the effect that the cargoes of subsistences "were purchased by certain civic officials in the name of the town, and then distributed in shares

[1] When a boat brought a cargo of coal to Würzburg, coal could only be sold in retail during the first eight days, each family being entitled to no more than fifty basketfuls. The remaining cargo could be sold wholesale, but the retailer was allowed to raise a *zittlicher* profit only, the *unzittlicher*, or dishonest profit, being strictly forbidden (Gramich, *l. c.*). Same in London (*Liber albus*, quoted by Ochenkowski, p. 161), and, in fact, everywhere.

[2] See Fagniez, *Études sur l'industrie et la classe industrielle à Paris au XIIIme et XIVme siècle*, Paris, 1877, pp. 155 *seq.* It hardly need be added that the tax on bread, and on beer as well, was settled after careful experiments as to the quantity of bread and beer which could be obtained from a given amount of corn. The Amiens archives contain the minutes of such experiences (A. de Calonne, *l. c.* pp. 77, 93). Also those of London (Ochenkowski, *England's wirthschaftliche Entwickelung, etc.*, Jena, 1879, p. 165).

among the merchant burgesses, no one being allowed to buy wares landed in the port unless the municipal authorities refused to purchase them. This seems—he adds—to have been quite a common practice in England, Ireland, Wales and Scotland."[1] Even in the sixteenth century we find that common purchases of corn were made for the "comoditie and profitt in all things of this . . . . Citie and Chamber of London, and of all the Citizens and Inhabitants of the same as moche as in us lieth"—as the Mayor wrote in 1565.[2] In Venice, the whole of the trade in corn is well known to have been in the hands of the city; the "quarters," on receiving the cereals from the board which administrated the imports, being bound to send to every citizen's house the quantity allotted to him.[3] In France, the city of Amiens used to purchase salt and to distribute it to all citizens at cost price;[4] and even now one sees in many French towns the *halles* which

1 Ch. Gross, *The Guild Merchant*, Oxford, 1890, i. 135. His documents prove that this practice existed in Liverpool (ii. 148–150), Waterford in Ireland, Neath in Wales, and Linlithgow and Thurso in Scotland. Mr. Gross's texts also show that the purchases were made for distribution, not only among the merchant burgesses, but "upon all citsains and commynalte" (p. 136, *note*), or, as the Thurso ordinance of the seventeenth century runs, to "make offer to the merchants, craftsmen, *and inhabitants* of the said burgh, that they may have their proportion of the same, according to their necessitys and ability."

2 *The Early History of the Guild of Merchant Taylors*, by Charles M. Clode, London, 1888, i. 361, appendix 10; also the following appendix which shows that the same purchases were made in 1546.

3 Cibrario, *Les conditions économiques de l'Italie au temps de Dante*, Paris, 1865, p. 44.

4 A. de Calonne, *La vie municipale au XVme siècle dans le Nord de la France*, Paris, 1880, pp. 12–16. In 1485 the city permitted the export to Antwerp of a certain quantity of corn, "the inhabitants of Antwerp being always ready to be agreeable to the merchants and burgesses of Amiens" (*ibid.*, pp. 75–77 and texts).

formerly were municipal *dépôts* for corn and salt.[1] In Russia it was a regular custom in Novgorod and Pskov.

The whole matter relative to the communal purchases for the use of the citizens, and the manner in which they used to be made, seems not to have yet received proper attention from the historians of the period; but there are here and there some very interesting facts which throw a new light upon it. Thus there is, among Mr. Gross's documents, a Kilkenny ordinance of the year 1367, from which we learn how the prices of the goods were established. "The merchants and the sailors," Mr. Gross writes, "were to state on oath the first cost of the goods and the expenses of transportation. Then the mayor of the town and two discreet men were to name the price at which the wares were to be sold." The same rule held good in Thurso for merchandise coming "by sea or land." This way of "naming the price" so well answers to the very conceptions of trade which were current in mediæval times that it must have been all but universal. To have the price established by a third person was a very old custom; and for all interchange within the city it certainly was a widely-spread habit to leave the establishment of prices to "discreet men"—to a third party—and not to the vendor or the buyer. But this order of things takes us still further back in the history of trade—namely, to a time when trade in staple produce was carried on by the whole city, and the merchants were only the commissioners, the trustees, of the city for selling the goods which it exported. A Waterford ordinance, published also by Mr. Gross, says "that all manere of marchandis *what*

[1] A. Babeau, *La ville sous l'ancien régime*, Paris, 1880.

so ever kynde thei be of . . . shal be bought by the Maire and balives which bene commene biers [common buyers, for the town] for the time being, and to distribute the same on freemen of the citie (the propre goods of free citisains and inhabitants only excepted)." This ordinance can hardly be explained otherwise than by admitting that all the exterior trade of the town was carried on by its agents. Moreover, we have direct evidence of such having been the case for Novgorod and Pskov. It was the Sovereign Novgorod and the Sovereign Pskov who sent their caravans of merchants to distant lands.

We know also that in nearly all mediæval cities of Middle and Western Europe, the craft guilds used to buy, as a body, all necessary raw produce, and to sell the produce of their work through their officials, and it is hardly possible that the same should not have been done for exterior trade—the more so as it is well known that up to the thirteenth century, not only all merchants of a given city were considered abroad as responsible in a body for debts contracted by any one of them, but the whole city as well was responsible for the debts of each one of its merchants. Only in the twelfth and thirteenth century the towns on the Rhine entered into special treaties abolishing this responsibility.[1] And finally we have the remarkable Ipswich document published by Mr. Gross, from which document we learn that the merchant guild of this town was constituted by all who had the freedom of the city, and who wished to pay their contribution ("their hanse") to the guild, the whole community discussing all together how better to maintain the merchant guild, and giving it certain privileges. The merchant guild of Ipswich

[1] Ennen, *Geschichte der Stadt Köln*, i. 491, 492, also texts.

thus appears rather as a body of trustees of the town than as a common private guild.

In short, the more we begin to know the mediæval city the more we see that it was not simply a political organization for the protection of certain political liberties. It was an attempt at organizing, on a much grander scale than in a village community, a close union for mutual aid and support, for consumption and production, and for social life altogether, without imposing upon men the fetters of the State, but giving full liberty of expression to the creative genius of each separate group of individuals in art, crafts, science, commerce, and political organization. How far this attempt has been successful will be best seen when we have analyzed in the next chapter the organization of labour in the mediæval city and the relations of the cities with the surrounding peasant population.

CHAPTER VI

MUTUAL AID IN THE MEDIÆVAL CITY (*continued*)

Likeness and diversity among the mediæval cities.—The craft-guilds: State-attributes in each of them.—Attitude of the city towards the peasants; attempts to free them.—The lords.—Results achieved by the mediæval city: in arts, in learning.—Causes of decay.

THE mediæval cities were not organized upon some preconceived plan in obedience to the will of an outside legislator. Each of them was a natural growth in the full sense of the word—an always varying result of struggle between various forces which adjusted and re-adjusted themselves in conformity with their relative energies, the chances of their conflicts, and the support they found in their surroundings. Therefore, there are not two cities whose inner organization and destinies would have been identical. Each one, taken separately, varies from century to century. And yet, when we cast a broad glance upon all the cities of Europe, the local and national unlikenesses disappear, and we are struck to find among all of them a wonderful resemblance, although each has developed for itself, independently from the others, and in different conditions. A small town in the north of Scotland, with its population of coarse labourers and fishermen; a rich city of Flanders, with its world-wide commerce, luxury, love of amusement and animated life; an Italian city enriched by its intercourse with the East, and breeding within its walls a refined artistic taste and civilization; and a poor, chiefly agricultural, city in the marsh and lake district of Russia, seem to have little in common. And nevertheless, the leading lines of their organization, and the spirit which animates them, are imbued with a strong family likeness. Everywhere we see the same federations of small communities and guilds, the same "sub-towns" round the mother city, the same folkmote, and the same insigns of its independence. The *defensor* of the city, under different names and in different accoutrements, represents the same authority and interests; food supplies, labour and commerce, are organized on closely similar lines; inner and outer struggles are fought with like ambitions; nay, the very formulæ used in the struggles, as also in the annals, the ordinances, and the rolls, are identical; and the architectural monuments, whether Gothic, Roman, or Byzantine in style, express the same aspirations and the same ideals; they are conceived and built in the same way. Many dissemblances are mere differences of age, and those disparities between sister cities which are real are repeated in different parts of Europe. The unity of the leading idea and the identity of origin make up for differences of climate, geographical situation, wealth, language and religion. This is why we can speak of *the* mediæval city as of a well-defined phase of civilization; and while every research insisting upon local and individual differences is most welcome, we may still indicate the chief lines of development which are common to all cities.[1]

[1] The literature of the subject is immense; but there is no work yet which treats of the mediæval city as of a whole. For the French Communes, Augustin Thierry's *Lettres* and *Considérations sur*

There is no doubt that the protection which used to be accorded to the market-place from the earliest barbarian times has played an important, though not an exclusive, part in the emancipation of the mediæval city. The early barbarians knew no trade within their village communities; they traded with strangers only, at certain definite spots, on certain determined days. And, in order that the stranger might come to the barter-place without risk of being slain for some feud which might be running between two kins, the market was always placed under the special protection of all kins. It was inviolable, like the place of worship under the shadow of which it was held. With the Kabyles it is still *annaya*, like the footpath along

---

l'histoire de France still remain classical, and Luchaire's *Communes françaises* is an excellent addition on the same lines. For the cities of Italy, the great work of Sismondi (*Histoire des républiques italiennes du moyen âge*, Paris, 1826, 16 vols.), Leo and Botta's *History of Italy*, Ferrari's *Révolutions d'Italie*, and Hegel's *Geschichte der Städteverfassung in Italien*, are the chief sources of general information. For Germany we have Maurer's *Städteverfassung*, Barthold's *Geschichte der deutschen Städte*, and, of recent works, Hegel's *Städte und Gilden der germanischen Völker* (2 vols. Leipzig, 1891), and Dr. Otto Kallsen's *Die deutschen Städte im Mittelalter* (2 vols. Halle, 1891), as also Janssen's *Geschichte des deutschen Volkes* (5 vols. 1886), which, let us hope, will soon be translated into English (French translation in 1892). For Belgium, A. Wauters, *Les Libertés communales* (Bruxelles, 1869–78, 3 vols.). For Russia, Byelaeff's, Kostomaroff's and Sergievich's works. And finally, for England, we posses one of the best works on cities of a wider region in Mrs. J. R. Green's *Town Life in the Fifteenth Century* (2 vols. London, 1894). We have, moreover, a wealth of well-known local histories, and several excellent works of general or economical history which I have so often mentioned in this and the preceding chapter. The richness of literature consists, however, chiefly in separate, sometimes admirable, researches into the history of separate cities, especially Italian and German; the guilds; the land question; the economical principles of the time; the economical importance of guilds and crafts; the leagues between cities (the Hansa); and communal art. An incredible wealth of information is contained in works of this second category, of which only some of the more important are named in these pages.

which women carry water from the well; neither must be trodden upon in arms, even during inter-tribal wars. In mediæval times the market universally enjoyed the same protection.[1] No feud could be prosecuted on the place whereto people came to trade, nor within a certain radius from it; and if a quarrel arose in the motley crowd of buyers and sellers, it had to be brought before those under whose protection the market stood—the community's tribunal, or the bishop's, the lord's, or the king's judge. A stranger who came to trade was a guest, and he went on under this very name. Even the lord who had no scruples about robbing a merchant on the high road, respected the *Weichbild*, that is, the pole which stood in the market-place and bore either the king's arms, or a glove, or the image of the local saint, or simply a cross, according to whether the market was under the protection of the king, the lord, the local church, or the folkmote—the *vyeche*.[2]

It is easy to understand how the self-jurisdiction of the city could develop out of the special jurisdiction in the market-place, when this last right was conceded, willingly or not, to the city itself. And such an origin

[1] Kulischer, in an excellent essay on primitive trade (*Zeitschrift für Völkerpsychologie*, Bd. x. 380), also points out that, according to Herodotus, the Argippæans were considered inviolable, because the trade between the Scythians and the northern tribes took place on their territory. A fugitive was sacred on their territory, and they were often asked to act as arbiters for their neighbours. See Appendix XI.

[2] Some discussion has lately taken place upon the *Weichbild* and the *Weichbild*-law, which still remain obscure (see Zöpfl, *Alterthümer des deutschen Reichs und Rechts*, iii. 29; Kallsen, i. 316). The above explanation seems to be the more probable, but, of course, it must be tested by further research. It is also evident that, to use a Scotch expression, the "mercet cross" could be considered as an emblem of Church jurisdiction, but we find it both in bishop cities and in those in which the folkmote was sovereign.

of the city's liberties, which can be traced in very many cases, necessarily laid a special stamp upon their subsequent development. It gave a predominance to the trading part of the community. The burghers who possessed a house in the city at the time being, and were co-owners in the town-lands, constituted very often a merchant guild which held in its hands the city's trade; and although at the outset every burgher, rich and poor, could make part of the merchant guild, and the trade itself seems to have been carried on for the entire city by its trustees, the guild gradually became a sort of privileged body. It jealously prevented the outsiders who soon began to flock into the free cities from entering the guild, and kept the advantages resulting from trade for the few "families" which had been burghers at the time of the emancipation. There evidently was a danger of a merchant oligarchy being thus constituted. But already in the tenth, and still more during the two next centuries, the chief crafts, also organized in guilds, were powerful enough to check the oligarchic tendencies of the merchants.

The craft guild was then a common seller of its produce and a common buyer of the raw materials, and its members were merchants and manual workers at the same time. Therefore, the predominance taken by the old craft guilds from the very beginnings of the free city life guaranteed to manual labour the high position which it afterwards occupied in the city.[1] In fact, in a

[1] For all concerning the merchant guild see Mr. Gross's exhaustive work, *The Guild Merchant* (Oxford, 1890, 2 vols.); also Mrs. Green's remarks in *Town Life in the Fifteenth Century*, vol. ii. chaps. v. viii. x.; and A. Doren's review of the subject in Schmoller's *Forschungen*, vol. xii. If the considerations indicated in the previous chapter (according to which trade was communal at its beginnings) prove to be correct, it will be permissible to suggest as a probable hypothesis that the guild merchant was a body entrusted

mediæval city manual labour was no token of inferiority; it bore, on the contrary, traces of the high respect it had been kept in in the village community. Manual labour in a "mystery" was considered as a pious duty towards the citizens: a public function (*Amt*), as honourable as any other. An idea of "justice" to the community, of "right" towards both producer and consumer, which would seem so extravagant now, penetrated production and exchange. The tanner's, the cooper's, or the shoemaker's work must be "just," fair, they wrote in those times. Wood, leather or thread which are used by the artisan must be "right"; bread must be baked "in justice," and so on. Transport this language into our present life, and it would seem affected and unnatural; but it was natural and unaffected then, because the mediæval artisan did not produce for an unknown buyer, or to throw his goods into an unknown market. He produced for his guild first; for a brotherhood of men who knew each other, knew the technics of the craft, and, in naming the price of each product, could appreciate the skill displayed in its fabrication or the labour bestowed upon it. Then the guild, not the separate producer, offered the goods for sale in the community, and this last, in its turn, offered to the brotherhood of allied communities those goods which were exported, and assumed responsibility for their

---

with commerce in the interest of the whole city, and only gradually became a guild of merchants trading for themselves; while the merchant adventurers of this country, the Novgorod *povolniki* (free colonizers and merchants) and the *mercati personati*, would be those to whom it was left to open new markets and new branches of commerce for themselves. Altogether, it must be remarked that the origin of the mediæval city can be ascribed to no separate agency. It was a result of *many* agencies in different degrees.

quality. With such an organization, it was the ambition of each craft not to offer goods of inferior quality, and technical defects or adulterations became a matter concerning the whole community, because, an ordinance says, "they would destroy public confidence."[1] Production being thus a social duty, placed under the control of the whole *amitas*, manual labour could not fall into the degraded condition which it occupies now, so long as the free city was living.

A difference between master and apprentice, or between master and worker (*compayne, Geselle*), existed in the mediæval cities from their very beginnings; but this was at the outset a mere difference of age and skill, not of wealth and power. After a seven years' apprenticeship, and after having proved his knowledge and capacities by a work of art, the apprentice became a master himself. And only much later, in the sixteenth century, after the royal power had destroyed the city and the craft organization, was it possible to become master in virtue of simple inheritance or wealth. But this was also the time of a general decay in mediæval industries and art.

There was not much room for hired work in the early flourishing periods of the mediæval cities, still less for individual hirelings. The work of the weavers, the archers, the smiths, the bakers, and so on, was performed for the craft and the city; and when craftsmen were hired in the building trades, they worked as temporary corporations (as they still do in the Russian *artéls*), whose work was paid *en bloc*. Work for a master began to multiply only later on; but even in this case the worker was paid better than he is paid now,

[1] Janssen's *Geschichte des deutschen Volkes*, i. 315; Gramich's *Würzburg*; and, in fact, any collection of ordinances.

even in this country, and very much better than he used to be paid all over Europe in the first half of this century. Thorold Rogers has familiarized English readers with this idea; but the same is true for the Continent as well, as is shown by the researches of Falke and Schönberg, and by many occasional indications. Even in the fifteenth century a mason, a carpenter, or a smith worker would be paid at Amiens four *sols* a day, which corresponded to forty-eight pounds of bread, or to the eighth part of a small ox (*bouvard*). In Saxony, the salary of the *Geselle* in the building trade was such that, to put it in Falke's words, he could buy with his six days' wages three sheep and one pair of shoes.[1] The donations of workers (*Geselle*) to cathedrals also bear testimony of their relative well-being, to say nothing of the glorious donations of certain craft guilds nor of what they used to spend in festivities and pageants.[2] In fact, the more we learn about the mediæval city, the more we are convinced that at no time has labour enjoyed such conditions of prosperity and such respect as when city life stood at its highest.

More than that; not only many aspirations of our

[1] Falke, *Geschichtliche Statistik*, i. 373–393, and ii. 66; quoted in Janssen's *Geschichte*, i. 339; J. D. Blavignac, in *Comptes et dépenses de la construction du clocher de Saint-Nicolas à Fribourg en Suisse*, comes to a similar conclusion. For Amiens, De Calonne's *Vie Municipale*, p. 99 and Appendix. For a thorough appreciation and graphical representation of the mediæval wages in England and their value in bread and meat, see G. Steffen's excellent article and curves in *The Nineteenth Century* for 1891, and *Studier öfver lönsystemets historia i England*, Stockholm, 1895.

[2] To quote but one example out of many which may be found in Schönberg's and Falke's works, the sixteen shoemaker workers (*Schusterknechte*) of the town Xanten, on the Rhine, gave, for erecting a screen and an altar in the church, 75 guldens of subscriptions, and 12 guldens out of their box, which money was worth, according to the best valuations, ten times its present value.

modern radicals were already realized in the middle ages, but much of what is described now as Utopian was accepted then as a matter of fact. We are laughed at when we say that work must be pleasant, but—"every one must be pleased with his work," a mediæval Kuttenberg ordinance says, "and no one shall, while doing nothing (*mit nichts thun*), appropriate for himself what others have produced by application and work, because laws must be a shield for application and work."[1] And amidst all present talk about an eight hours' day, it may be well to remember an ordinance of Ferdinand the First relative to the Imperial coal mines, which settled the miner's day at eight hours, "as it used to be of old" (*wie vor Alters herkommen*), and work on Saturday afternoon was prohibited. Longer hours were very rare, we are told by Janssen, while shorter hours were of common occurrence. In this country, in the fifteenth century, Rogers says, "the workmen worked only forty-eight hours a week."[2] The Saturday half-holiday, too, which we consider as a modern conquest, was in reality an old mediæval institution; it was bathing-time for a great part of the community, while Wednesday afternoon was bathing-time for the *Geselle*.[3] And although school meals did

[1] Quoted by Janssen, *l. c.* i. 343.
[2] *The Economical Interpretation of History*, London, 1891, p. 303.
[3] Janssen, *l. c.* See also Dr. Alwin Schultz, *Deutsches Leben im XIV. und XV. Jahrhundert*, grosse Ausgabe, Wien, 1892, pp. 67 *seq.* At Paris, the day of labour varied from seven to eight hours in the winter to fourteen hours in summer in certain trades, while in others it was from eight to nine hours in winter, to from ten to twelve in summer. All work was stopped on Saturdays and on about twenty-five other days (*jours de commun de vile foire*) at four o'clock, while on Sundays and thirty other holidays there was no work at all. The general conclusion is, that the mediæval worker worked *less* hours, all taken, than the present-day worker (Dr. E. Martin Saint-Léon, *Histoire des corporations*, p. 121).

not exist—probably because no children went hungry to school—a distribution of bath-money to the children whose parents found difficulty in providing it was habitual in several places. As to Labour Congresses, they also were a regular feature of the middles ages. In some parts of Germany craftsmen of the same trade, belonging to different communes, used to come together every year to discuss questions relative to their trade, the years of apprenticeship, the wandering years, the wages, and so on; and in 1572, the Hanseatic towns formally recognized the right of the crafts to come together at periodical congresses, and to take any resolutions, so long as they were not contrary to the cities' rolls, relative to the quality of goods. Such Labour Congresses, partly international like the Hansa itself, are known to have been held by bakers, founders, smiths, tanners, sword-makers and cask-makers.[1]

The craft organization required, of course, a close supervision of the craftsmen by the guild, and special jurates were always nominated for that purpose. But it is most remarkable that, so long as the cities lived their free life, no complaints were heard about the supervision; while, after the State had stepped in, confiscating the property of the guilds and destroying their independence in favour of its own bureaucracy, the complaints became simply countless.[2] On the other hand, the immensity of progress realized in all

[1] W. Stieda, "Hansische Vereinbarungen über städtisches Gewerbe im XIV. und XV. Jahrhundert," in *Hansische Geschichtsblätter*, Jahrgang 1886, p. 121. Schönberg's *Wirthschaftliche Bedeutung der Zünfte;* also, partly, Roscher.

[2] See Toulmin Smith's deeply-felt remarks about the royal spoliation of the guilds, in Miss Smith's Introduction to *English Guilds*. In France the same royal spoliation and abolition of the guilds' jurisdiction was begun from 1306, and the final blow was struck in 1382 (Fagniez, *l. c.* pp. 52–54).

arts under the mediæval guild system is the best proof that the system was no hindrance to individual initiative.[1] The fact is, that the mediæval guild, like the mediæval parish, "street," or "quarter," was not a body of citizens, placed under the control of State functionaries; it was a union of all men connected with a given trade: jurate buyers of raw produce, sellers of manufactured goods, and artisans—masters, "compaynes," and apprentices. For the inner organization of the trade its assembly was sovereign, so long as it did not hamper the other guilds, in which case the matter was brought before the guild of the guilds—the city. But there was in it something more than that. It had its own self-jurisdiction, its own military force, its own general assemblies, its own traditions of struggles, glory, and independence, its own relations with other guilds of the same trade in other cities: it had, in a word, a full organic life which could only result from the integrality of the vital functions. When the town was called to arms, the guild appeared as a separate company (*Schaar*), armed with its own arms (or its own guns, lovingly decorated by the guild, at a subsequent epoch), under its own self-elected commanders. It was, in a word, as independent a unit of the federation as the republic of Uri or Geneva was fifty years ago in the Swiss Confederation. So that,

[1] Adam Smith and his contemporaries knew well what they were condemning when they wrote against the *State* interference in trade and the trade monopolies of *State* creation. Unhappily, their followers, with their hopeless superficiality, flung mediæval guilds and State interference into the same sack, making no distinction between a Versailles edict and a guild ordinance. It hardly need be said that the economists who have seriously studied the subject, like Schönberg (the editor of the well-known course of *Political Economy*), never fell into such an error. But, till lately, diffuse discussions of the above type went on for economical "science."

to compare it with a modern trade union, divested of all attributes of State sovereignty, and reduced to a couple of functions of secondary importance, is as unreasonable as to compare Florence or Brügge with a French commune vegetating under the Code Napoléon, or with a Russian town placed under Catherine the Second's municipal law. Both have elected mayors, and the latter has also its craft corporations; but the difference is—all the difference that exists between Florence and Fontenay-les-Oies or Tsarevokokshaisk, or between a Venetian doge and a modern mayor who lifts his hat before the *sous-préfet's* clerk.

The mediæval guilds were capable of maintaining their independence; and, later on, especially in the fourteenth century, when, in consequence of several causes which shall presently be indicated, the old municipal life underwent a deep modification, the younger crafts proved strong enough to conquer their due share in the management of the city affairs. The masses, organized in "minor" arts, rose to wrest the power out of the hands of a growing oligarchy, and mostly succeeded in this task, opening again a new era of prosperity. True, that in some cities the uprising was crushed in blood, and mass decapitations of workers followed, as was the case in Paris in 1306, and in Cologne in 1371. In such cases the city's liberties rapidly fell into decay, and the city was gradually subdued by the central authority. But the majority of the towns had preserved enough of vitality to come out of the turmoil with a new life and vigour.[1] A new

[1] In Florence the seven minor arts made their revolution in 1270–82, and its results are fully described by Perrens (*Histoire de Florence*, Paris, 1877, 3 vols.), and especially by Gino Capponi (*Storia della*

period of rejuvenescence was their reward. New life was infused, and it found its expression in splendid architectural monuments, in a new period of prosperity, in a sudden progress of technics and invention, and in a new intellectual movement leading to the Renaissance and to the Reformation.

The life of a mediæval city was a succession of hard battles to conquer liberty and to maintain it. True, that a strong and tenacious race of burghers had developed during those fierce contests; true, that love and worship of the mother city had been bred by these struggles, and that the grand things achieved by the mediæval communes were a direct outcome of that love. But the sacrifices which the communes had to sustain in the battle for freedom were, nevertheless, cruel, and left deep traces of division on their inner life as well. Very few cities had succeeded, under a concurrence of favourable circumstances, in obtaining

---

repubblica di Firenze, 2da edizione, 1876, i. 58–80; translated into German). In Lyons, on the contrary, where the movement of the minor crafts took place in 1402, the latter were defeated and lost the right of themselves nominating their own judges. The two parties came apparently to a compromise. In Rostock the same movement took place in 1313; in Zürich in 1336; in Bern in 1363; in Braunschweig in 1374, and next year in Hamburg; in Lübeck in 1376–84; and so on. See Schmoller's *Strassburg zur Zeit der Zunftkämpfe* and *Strassburg's Blüthe*; Brentano's *Arbeitergilden der Gegenwart*, 2 vols., Leipzig, 1871–72; Eb. Bain's *Merchant and Craft Guilds*, Aberdeen, 1887, pp. 26–47, 75, etc. As to Mr. Gross's opinion relative to the same struggles in England, see Mrs. Green's remarks in her *Town Life in the Fifteenth Century*, ii. 190–217; also the chapter on the Labour Question, and, in fact, the whole of this extremely interesting volume. Brentano's views on the crafts' struggles, expressed especially in §§ iii. and iv. of his essay "On the History and Development of Guilds," in Toulmin Smith's *English Guilds* remain classical for the subject, and may be said to have been again and again confirmed by subsequent research.

liberty at one stroke, and these few mostly lost it equally easily; while the great number had to fight fifty or a hundred years in succession, often more, before their rights to free life had been recognized, and another hundred years to found their liberty on a firm basis—the twelfth century charters thus being but one of the stepping-stones to freedom.[1] In reality, the mediæval city was a fortified oasis amidst a country plunged into feudal submission, and it had to make room for itself by the force of its arms. In consequence of the causes briefly alluded to in the preceding chapter, each village community had gradually fallen under the yoke of some lay or clerical lord. His house had grown to be a castle, and his brothers-in-arms were now the scum of adventurers, always ready to plunder the peasants. In addition to three days a week which the peasants had to work for the lord, they had also to bear all sorts of exactions for the right to sow and to crop, to be gay or sad, to live, to marry, or to die. And, worst of all, they were continually plundered by the armed robbers of some neighbouring lord, who chose to consider them as their master's kin, and to take upon them, and upon their cattle and crops, the revenge for a feud he was fighting against their owner. Every meadow, every field, every river, and road around the city, and every man upon the land was under some lord.

The hatred of the burghers towards the feudal barons has found a most characteristic expression in

[1] To give but one example—Cambrai made its first revolution in 907, and, after three or four more revolts, it obtained its charter in 1076. This charter was repealed twice (1107 and 1138), and twice obtained again (in 1127 and 1180). Total, 223 years of struggles before conquering the right to independence. Lyons—from 1195 to 1320.

the wording of the different charters which they compelled them to sign. Heinrich V. is made to sign in the charter granted to Speier in 1111, that he frees the burghers from "the horrible and execrable law of mortmain, through which the town has been sunk into deepest poverty" (*von dem scheusslichen und nichtswürdigen Gesetze, welches gemein Budel genannt wird,* Kallsen, i. 307). The *coutume* of Bayonne, written about 1273, contains such passages as these: "The people is anterior to the lords. It is the people, more numerous than all others, who, desirous of peace, has made the lords for bridling and knocking down the powerful ones," and so on (Giry, *Établissements de Rouen*, i. 117, quoted by Luchaire, p. 24). A charter submitted for King Robert's signature is equally characteristic. He is made to say in it: "I shall rob no oxen nor other animals. I shall seize no merchants, nor take their moneys, nor impose ransom. From Lady Day to the All Saints' Day I shall seize no horse, nor mare, nor foals, in the meadows. I shall not burn the mills, nor rob the flour. . . . I shall offer no protection to thieves," etc. (Pfister has published that document, reproduced by Luchaire). The charter "granted" by the Besançon Archbishop Hugues, in which he has been compelled to enumerate all the mischiefs due to his mortmain rights, is equally characteristic.[1] And so on.

Freedom could not be maintained in such surroundings, and the cities were compelled to carry on the war outside their walls. The burghers sent out emissaries to lead revolt in the villages; they received villages

[1] See Tuetey, "Étude sur le droit municipal . . . en Franche-Comté," in *Mémoires de la Société d'émulation de Montbéliard*, 2e série, ii. 129 *seq.*

into their corporations, and they waged direct war against the nobles. It Italy, where the land was thickly sprinkled with feudal castles, the war assumed heroic proportions, and was fought with a stern acrimony on both sides. Florence sustained for seventy-seven years a succession of bloody wars, in order to free its *contado* from the nobles; but when the conquest had been accomplished (in 1181) all had to begin anew. The nobles rallied; they constituted their own leagues in opposition to the leagues of the towns, and, receiving fresh support from either the Emperor or the Pope, they made the war last for another 130 years. The same took place in Rome, in Lombardy, all over Italy.

Prodigies of valour, audacity, and tenaciousness were displayed by the citizens in these wars. But the bows and the hatchets of the arts and crafts had not always the upper hand in their encounters with the armour-clad knights, and many castles withstood the ingenious siege-machinery and the perseverance of the citizens. Some cities, like Florence, Bologna, and many towns in France, Germany, and Bohemia, succeeded in emancipating the surrounding villages, and they were rewarded for their efforts by an extraordinary prosperity and tranquillity. But even here, and still more in the less strong or less impulsive towns, the merchants and artisans, exhausted by war, and misunderstanding their own interests, bargained over the peasants' heads. They compelled the lord to swear allegiance to the city; his country castle was dismantled, and he agreed to build a house and to reside in the city, of which he became a co-burgher (*com-bourgeois, con-cittadino*); but he maintained in return most of his rights upon the peasants, who only won a partial relief from their burdens. The burgher could not understand that equal rights of citizenship might be granted to the peasant upon whose food supplies he had to rely, and a deep rent was traced between town and village. In some cases the peasants simply changed owners, the city buying out the barons' rights and selling them in shares to her own citizens.[1] Serfdom was maintained, and only much later on, towards the end of the thirteenth century, it was the craft revolution which undertook to put an end to it, and abolished personal servitude, but dispossessed at the same time the serfs of the land.[2] It hardly need be added that the fatal results of such policy were soon felt by the cities themselves; the country became the city's enemy.

The war against the castles had another bad effect. It involved the cities in a long succession of mutual wars, which have given origin to the theory, till lately in vogue, namely, that the towns lost their independence through their own jealousies and mutual fights. The imperialist historians have especially supported this theory, which, however, is very much undermined now by modern research. It is certain that in Italy cities fought each other with a stubborn animosity, but nowhere else did such contests attain the same proportions; and in Italy itself the city wars,

[1] This seems to have been often the case in Italy. In Switzerland, Bern bought even the towns of Thun and Burgdorf.

[2] Such was, at least, the case in the cities of Tuscany (Florence, Lucca, Sienna, Bologna, etc.), for which the relations between city and peasants are best known. (Luchitzkiy, "Slavery and Russian Slaves in Florence," in Kieff University *Izvestia* for 1885, who has perused Rumohr's *Ursprung der Besitzlosigkeit der Colonien in Toscana*, 1830.) The whole matter concerning the relations between the cities and the peasants requires much more study than has hitherto been done

especially those of the earlier period, had their special causes. They were (as was already shown by Sismondi and Ferrari) a mere continuation of the war against the castles—the free municipal and federative principle unavoidably entering into a fierce contest with feudalism, imperialism, and papacy. Many towns which had but partially shaken off the yoke of the bishop, the lord, or the Emperor, were simply driven against the free cities by the nobles, the Emperor, and Church, whose policy was to divide the cities and to arm them against each other. These special circumstances (partly reflected on to Germany also) explain why the Italian towns, some of which sought support with the Emperor to combat the Pope, while the others sought support from the Church to resist the Emperor, were soon divided into a Gibelin and a Guelf camp, and why the same division appeared in each separate city.[1]

The immense economical progress realized by most Italian cities just at the time when these wars were hottest,[2] and the alliances so easily concluded between towns, still better characterize those struggles and further undermine the above theory. Already in the years 1130–1150 powerful leagues came into existence; and a few years later, when Frederick Barbarossa invaded Italy and, supported by the nobles and some retardatory cities, marched against Milan, popular enthusiasm was roused in many towns by popular

[1] Ferrari's generalizations are often too theoretical to be always correct; but his views upon the part played by the nobles in the city wars are based upon a wide range of authenticated facts.

[2] Only such cities as stubbornly kept to the cause of the barons, like Pisa or Verona, lost through the wars. For many towns which fought on the barons' side, the defeat was also the beginning of liberation and progress.

preachers. Crema, Piacenza, Brescia, Tortona, etc., went to the rescue; the banners of the guilds of Verona, Padua, Vicenza, and Trevisa floated side by side in the cities' camp against the banners of the Emperor and the nobles. Next year the Lombardian League came into existence, and sixty years later we see it reinforced by many other cities, and forming a lasting organization which had half of its federal war-chest in Genoa and the other half in Venice.[1] In Tuscany, Florence headed another powerful league, to which Lucca, Bologna, Pistoia, etc., belonged, and which played an important part in crushing down the nobles in middle Italy, while smaller leagues were of common occurrence. It is thus certain that although petty jealousies undoubtedly existed, and discord could be easily sown, they did not prevent the towns from uniting together for the common defence of liberty. Only later on, when separate cities became little States, wars broke out between them, as always must be the case when States struggle for supremacy or colonies.

Similar leagues were formed in Germany for the same purpose. When, under the successors of Conrad, the land was the prey of interminable feuds between the nobles, the Westphalian towns concluded a league against the knights, one of the clauses of which was never to lend money to a knight who would continue to conceal stolen goods.[2] When "the knights and the nobles lived on plunder, and murdered whom they chose to murder," as the *Wormser Zorn* complains, the cities on the Rhine (Mainz, Cologne, Speier, Strasburg, and Basel) took the initiative of a league which soon

[1] Ferrari, ii. 18, 104 *seq.*; Leo and Botta, i. 432.

[2] Joh. Falke, *Die Hansa als Deutsche See- und Handelsmacht*, Berlin, 1863, pp. 31, 55.

numbered sixty allied towns, repressed the robbers, and maintained peace. Later on, the league of the towns of Suabia, divided into three "peace districts" (Augsburg, Constance, and Ulm), had the same purpose. And even when such leagues were broken,[1] they lived long enough to show that while the supposed peacemakers—the kings, the emperors, and the Church—fomented discord, and were themselves helpless against the robber knights, it was from the cities that the impulse came for re-establishing peace and union. The cities—not the emperors—were the real makers of the national unity.[2]

Similar federations were organized for the same purpose among small villages, and now that attention has been drawn to this subject by Luchaire we may expect soon to learn much more about them. Villages joined into small federations in the *contado* of Florence, so also in the dependencies of Novgorod and Pskov. As to France, there is positive evidence of a federation of seventeen peasant villages which has existed in the Laonnais for nearly a hundred years (till 1256), and has fought hard for its independence. Three more peasant republics, which had sworn charters similar to those of Laon and Soissons, existed in the neighbourhood of Laon, and, their territories being contiguous, they supported each other in their liberation wars. Altogether, Luchaire is of the opinion that many such federations must have come into existence in France in the twelfth and thirteenth centuries, but that documents relative to them are mostly lost. Of course,

[1] For Aachen and Cologne we have direct testimony that the bishops of these two cities—one of them bought by the enemy—opened to him the gates.

[2] See the facts, though not always the conclusions, of Nitzsch, iii. 133 *seq.*; also Kallsen, i. 458, etc.

being unprotected by walls, they could easily be crushed down by the kings and the lords; but in certain favourable circumstances, when they found support in a league of towns and protection in their mountains, such peasant republics became independent units of the Swiss Confederation.[1]

As to unions between cities for peaceful purposes, they were of quite common occurrence. The intercourse which had been established during the period of liberation was not interrupted afterwards. Sometimes, when the *scabini* of a German town, having to pronounce judgment in a new or complicated case, declared that they knew not the sentence (*des Urtheiles nicht weise zu sein*), they sent delegates to another city to get the sentence. The same happened also in France;[2] while Forli and Ravenna are known to have mutually naturalized their citizens and granted them full rights in both cities. To submit a contest arisen between two towns, or within a city, to another commune which was invited to act as arbiter, was also in the spirit of the times.[3] As to commercial treaties between cities, they were quite habitual.[4] Unions for regulating the production and the sizes of casks which were used for

[1] On the Commune of the Laonnais, which, until Melleville's researches (*Histoire de la Commune du Laonnais*, Paris, 1853), was confounded with the Commune of Laon, see Luchaire, pp. 75 *seq.* For the early peasants' guilds and subsequent unions see R. Wilman's "Die ländlichen Schutzgilden Westphaliens," in *Zeitschrift für Kulturgeschichte*, neue Folge, Bd. iii., quoted in Henne-am-Rhyn's *Kulturgeschichte*, iii. 249.

[2] Luchaire, p. 149.

[3] Two important cities, like Mainz and Worms, would settle a political contest by means of arbitration. After a civil war broken out in Abbeville, Amiens would act, in 1231, as arbiter (Luchaire, 149); and so on.

[4] See, for instance, W. Stieda, *Hansische Vereinbarungen*, *l. c.*, p. 114.

the commerce in wine, "herring unions," and so on, were mere precursors of the great commercial federations of the Flemish Hansa, and, later on, of the great North German Hansa, the history of which alone might contribute pages and pages to illustrate the federation spirit which permeated men at that time. It hardly need be added, that through the Hanseatic unions the mediæval cities have contributed more to the development of international intercourse, navigation, and maritime discovery than all the States of the first seventeen centuries of our era.

In a word, federations between small territorial units, as well as among men united by common pursuits within their respective guilds, and federations between cities and groups of cities constituted the very essence of life and thought during that period. The first five of the second decade of centuries of our era may thus be described as an immense attempt at securing mutual aid and support on a grand scale, by means of the principles of federation and association carried on through all manifestations of human life and to all possible degrees. This attempt was attended with success to a very great extent. It united men formerly divided; it secured them a very great deal of freedom, and it tenfolded their forces. At a time when particularism was bred by so many agencies, and the causes of discord and jealousy might have been so numerous, it is gratifying to see that cities scattered over a wide continent had so much in common, and were so ready to confederate for the prosecution of so many common aims. They succumbed in the long run before powerful enemies; not having understood the mutual-aid principle widely enough, they themselves committed fatal faults; but they did not perish through their own jealousies, and their errors were not a want of federation spirit among themselves.

The results of that new move which mankind made in the mediæval city were immense. At the beginning of the eleventh century the towns of Europe were small clusters of miserable huts, adorned but with low clumsy churches, the builders of which hardly knew how to make an arch; the arts, mostly consisting of some weaving and forging, were in their infancy; learning was found in but a few monasteries. Three hundred and fifty years later, the very face of Europe had been changed. The land was dotted with rich cities, surrounded by immense thick walls which were embellished by towers and gates, each of them a work of art in itself. The cathedrals, conceived in a grand style and profusely decorated, lifted their bell-towers to the skies, displaying a purity of form and a boldness of imagination which we now vainly strive to attain. The crafts and arts had risen to a degree of perfection which we can hardly boast of having superseded in many directions, if the inventive skill of the worker and the superior finish of his work be appreciated higher than rapidity of fabrication. The navies of the free cities furrowed in all directions the Northern and the Southern Mediterranean; one effort more, and they would cross the oceans. Over large tracts of land well-being had taken the place of misery; learning had grown and spread. The methods of science had been elaborated; the basis of natural philosophy had been laid down; and the way had been paved for all the mechanical inventions of which our own times are so proud. Such were the magic changes accomplished in Europe in less than four hundred years. And the losses which Europe sustained through the loss of its free cities can only be understood when we compare the seventeenth century with the fourteenth or the thirteenth. The prosperity which formerly characterized Scotland, Germany, the plains of Italy, was gone. The roads had fallen into an abject state, the cities were depopulated, labour was brought into slavery, art had vanished, commerce itself was decaying.[1]

If the mediæval cities had bequeathed to us no written documents to testify of their splendour, and left nothing behind but the monuments of building art which we see now all over Europe, from Scotland to Italy, and from Gerona in Spain to Breslau in Slavonian territory, we might yet conclude that the times of independent city life were times of the greatest development of human intellect during the Christian era down to the end of the eighteenth century. On looking, for instance, at a mediæval picture representing Nuremberg with its scores of towers and lofty spires, each of which bore the stamp of free creative art, we can hardly conceive that three hundred years before the town was but a collection of miserable hovels. And our admiration grows when we go into the details of the architecture and decorations of each of the countless churches, bell-towers, gates, and communal houses which are scattered all over Europe as far east as Bohemia and the now dead towns of Polish Galicia. Not only Italy, that mother of art, but all

[1] Cosmo Innes's *Early Scottish History* and *Scotland in Middle Ages*, quoted by Rev. Denton, *l.c.*, pp. 68, 69; Lamprecht's *Deutsches wirthschaftliche Leben im Mittelalter*, review by Schmoller in his *Jahrbuch*, Bd. xii.; Sismondi's *Tableau de l'agriculture toscane*, pp. 226 *seq.* The dominions of Florence could be recognized at a glance through their prosperity.

Europe is full of such monuments. The very fact that of all arts architecture—a social art above all—had attained the highest development, is significant in itself. To be what it was, it must have originated from an eminently social life.

Mediæval architecture attained its grandeur—not only because it was a natural development of handicraft; not only because each building, each architectural decoration, had been devised by men who knew through the experience of their own hands what artistic effects can be obtained from stone, iron, bronze, or even from simple logs and mortar; not only because each monument was a result of collective experience, accumulated in each "mystery" or craft[1]—it was grand because it was born out of a grand idea. Like Greek art, it sprang out of a conception of brotherhood and unity fostered by the city. It had an audacity which could only be won by audacious struggles and victories; it had that expression of vigour, because vigour permeated all the life of the city. A cathedral or a communal house symbolized the grandeur of an organism of which every mason and stone-cutter was the builder, and a mediæval building appears—not as a solitary effort to which

[1] Mr. John J. Ennett (*Six Essays*, London, 1891) has excellent pages on this aspect of mediæval architecture. Mr. Willis, in his appendix to Whewell's *History of Inductive Sciences* (i. 261–262), has pointed out the beauty of the mechanical relations in mediæval buildings. "A new decorative construction was matured," he writes, "not thwarting and controlling, but assisting and harmonizing with the mechanical construction. Every member, every moulding, becomes a sustainer of weight; and by the multiplicity of props assisting each other, and the consequent subdivision of weight, the eye was satisfied of the stability of the structure, notwithstanding curiously slender aspects of the separate parts." An art which sprang out of the *social* life of the city could not be better characterized.

thousands of slaves would have contributed the share assigned them by one man's imagination; all the city contributed to it. The lofty bell-tower rose upon a structure, grand in itself, in which the life of the city was throbbing—not upon a meaningless scaffold like the Paris iron tower, not as a sham structure in stone intended to conceal the ugliness of an iron frame, as has been done in the Tower Bridge. Like the Acropolis of Athens, the cathedral of a mediæval city was intended to glorify the grandeur of the victorious city, to symbolize the union of its crafts, to express the glory of each citizen in a city of his own creation. After having achieved its craft revolution, the city often began a new cathedral in order to express the new, wider, and broader union which had been called into life.

The means at hand for these grand undertakings were disproportionately small. Cologne Cathedral was begun with a yearly outlay of but 500 marks; a gift of 100 marks was inscribed as a grand donation;[1] and even when the work approached completion, and gifts poured in in proportion, the yearly outlay in money stood at about 5,000 marks, and never exceeded 14,000. The cathedral of Basel was built with equally small means. But each corporation contributed its part of stone, work, and decorative genius to *their* common monument. Each guild expressed in it its political conceptions, telling in stone or in bronze the history of the city, glorifying the principles of "Liberty, equality, and fraternity,"[2] praising the city's allies, and sending to eternal fire its enemies.

[1] Dr. L. Ennen, *Der Dom zu Köln, seine Construction und Anstaltung*, Kòln, 1871.

[2] The three statues are among the outer decorations of Nôtre Dame de Paris.

And each guild bestowed its *love* upon the communal monument by richly decorating it with stained windows, paintings, "gates, worthy to be the gates of Paradise," as Michel Angelo said, or stone decorations of each minutest corner of the building.[1] Small cities, even small parishes,[2] vied with the big agglomerations in this work, and the cathedrals of Laon and St. Ouen hardly stand behind that of Rheims, or the Communal House of Bremen, or the folkmote's bell-tower of Breslau. "No works must be begun by the commune but such as are conceived in response to the grand heart of the commune, composed of the hearts of all citizens, united in one common will"—such were the words of the Council of Florence; and this spirit appears in all communal works of common utility, such as the canals, terraces, vineyards, and fruit gardens around Florence, or the irrigation canals which intersected the plains of Lombardy, or the port and aqueduct of Genoa, or, in fact, any works of the kind which were achieved by almost every city.[3]

All arts had progressed in the same way in the mediæval cities, those of our own days mostly being but a continuation of what had grown at that time. The prosperity of the Flemish cities was based upon the fine woollen cloth they fabricated. Florence, at the

[1] Mediæval art, like Greek art, did not know those curiosity-shops which we call a National Gallery or a Museum. A picture was painted, a statue was carved, a bronze decoration was cast to stand in its proper place in a monument of communal art. It lived there, it was part of a whole, and it contributed to give unity to the impression produced by the whole.

[2] Cf. J. T. Ennett's "Second Essay," p. 36.

[3] Sismondi, iv. 172; xvi. 356. The great canal, *Naviglio Grande*, which brings the water from the Tessino, was begun in 1179, *i. e.* after the conquest of independence, and it was ended in the thirteenth century. On the subsequent decay, see xvi. 355.

beginning of the fourteenth century, before the black death, fabricated from 70,000 to 100,000 *panni* of woollen stuffs, which were valued at 1,200,000 golden florins.[1] The chiselling of precious metals, the art of casting, the fine forging of iron, were creations of the mediæval "mysteries" which had succeeded in attaining in their own domains all that could be made by the hand, without the use of a powerful prime motor. By the hand and by invention, because, to use Whewell's words:

> "Parchment and paper, printing and engraving, improved glass and steel, gunpowder, clocks, telescopes, the mariner's compass, the reformed calendar, the decimal notation; algebra, trigonometry, chemistry, counterpoint (an invention equivalent to a new creation of music); these are all possessions which we inherit from that which has so disparagingly been termed the Stationary Period" (*History of Inductive Sciences*, i. 252).

True that no new principle was illustrated by any of these discoveries, as Whewell said; but mediæval science had done something more than the actual discovery of new principles. It had prepared the discovery of all the new principles which we know at the present time in mechanical sciences: it had accustomed the explorer to observe facts and to reason from them. It was inductive science, even though it had not yet fully grasped the importance and the powers of induction; and it laid the foundations of

[1] In 1336 it had 8,000 to 10,000 boys and girls in its primary schools, 1,000 to 1,200 boys in its seven middle schools, and from 550 to 600 students in its four universities. The thirty communal hospitals contained over 1,000 beds for a population of 90,000 inhabitants (Capponi, ii. 249 *seq.*). It has more than once been suggested by authoritative writers that education stood, as a rule, at a much higher level than is generally supposed. Certainly so in democratic Nuremberg.

both mechanics and natural philosophy. Francis Bacon, Galileo, and Copernicus were the direct descendants of a Roger Bacon and a Michael Scot, as the steam engine was a direct product of the researches carried on in the Italian universities on the weight of the atmosphere, and of the mathematical and technical learning which characterized Nuremberg.

But why should one take trouble to insist upon the advance of science and art in the mediæval city? Is it not enough to point to the cathedrals in the domain of skill, and to the Italian language and the poem of Dante in the domain of thought, to give at once the measure of what the mediæval city *created* during the four centuries it lived?

The mediæval cities have undoubtedly rendered an immense service to European civilization. They have prevented it from being drifted into the theocracies and despotical states of old; they have endowed it with the variety, the self-reliance, the force of initiative, and the immense intellectual and material energies it now possesses, which are the best pledge for its being able to resist any new invasion of the East. But why did these centres of civilization, which attempted to answer to deeply-seated needs of human nature, and were so full of life, not live further on? Why were they seized with senile debility in the sixteenth century? and, after having repulsed so many assaults from without, and only borrowed new vigour from their interior struggles, why did they finally succumb to both?

Various causes contributed to this effect, some of them having their roots in the remote past, while others originated in the mistakes committed by the cities themselves. Towards the end of the fifteenth century, mighty States, reconstructed on the old Roman pattern, were already coming into existence. In each country and each region some feudal lord, more cunning, more given to hoarding, and often less scrupulous than his neighbours, had succeeded in appropriating to himself richer personal domains, more peasants on his lands, more knights in his following, more treasures in his chest. He had chosen for his seat a group of happily-situated villages, not yet trained into free municipal life—Paris, Madrid, or Moscow—and with the labour of his serfs he had made of them royal fortified cities, whereto he attracted war companions by a free distribution of villages, and merchants by the protection he offered to trade. The germ of a future State, which began gradually to absorb other similar centres, was thus laid. Lawyers, versed in the study of Roman law, flocked into such centres; a tenacious and ambitious race of men issued from among the burgesses, who equally hated the naughtiness of the lords and what they called the lawlessness of the peasants. The very forms of the village community, unknown to their code, the very principles of federalism were repulsive to them as "barbarian" inheritances. Cæsarism, supported by the fiction of popular consent and by the force of arms, was their ideal, and they worked hard for those who promised to realize it.[1]

[1] Cf. L. Ranke's excellent considerations upon the essence of Roman law in his *Weltgeschichte*, Bd. iv. Abth. 2, pp. 20–31. Also Sismondi's remarks upon the part played by the *légistes* in the constitution of royal authority, *Histoire des Français*, Paris, 1826, viii. 85–99. The popular hatred against these "*weise Doktoren und Beutelschneider des Volks*" broke out with full force in the first years of the sixteenth century in the sermons of the early Reform movement.

The Christian Church, once a rebel against Roman law and now its ally, worked in the same direction. The attempt at constituting the theocratic Empire of Europe having proved a failure, the more intelligent and ambitious bishops now yielded support to those whom they reckoned upon for reconstituting the power of the Kings of Israel or of the Emperors of Constantinople. The Church bestowed upon the rising rulers her sanctity, she crowned them as God's representatives on earth, she brought to their service the learning and the statesmanship of her ministers, her blessings and maledictions, her riches, and the sympathies she had retained among the poor. The peasants, whom the cities had failed or refused to free, on seeing the burghers impotent to put an end to the interminable wars between the knights—which wars they had so dearly to pay for—now set their hopes upon the King, the Emperor, or the Great Prince; and while aiding them to crush down the mighty feudal owners, they aided them to constitute the centralized State. And finally, the invasions of the Mongols and the Turks, the holy war against the Maures in Spain, as well as the terrible wars which soon broke out between the growing centres of sovereignty—Ile de France and Burgundy, Scotland and England, England and France, Lithuania and Poland, Moscow and Tver, and so on—contributed to the same end. Mighty States made their appearance; and the cities had now to resist not only loose federations of lords, but strongly-organized centres, which had armies of serfs at their disposal.

The worst was, that the growing autocracies found support in the divisions which had grown within the cities themselves. The fundamental idea of the mediæval city was grand, but it was not wide enough. Mutual aid and support cannot be limited to a small association; they must spread to its surroundings, or else the surroundings will absorb the association. And in this respect the mediæval citizen had committed a formidable mistake at the outset. Instead of looking upon the peasants and artisans who gathered under the protection of his walls as upon so many aids who would contribute their part to the making of the city—as they really did—a sharp division was traced between the "families" of old burghers and the new-comers. For the former, all benefits from communal trade and communal lands were reserved, and nothing was left for the latter but the right of freely using the skill of their own hands. The city thus became divided into "the burghers" or "the commonalty," and "the inhabitants."[1] The trade, which was formerly communal, now became the privilege of the merchant and artisan "families," and the next step—that of becoming individual, or the privilege of oppressive trusts—was unavoidable.

The same division took place between the city proper and the surrounding villages. The commune had well tried to free the peasants, but her wars against the lords became, as already mentioned, wars for freeing the city itself from the lords, rather than for freeing the peasants. She left to the lord his rights over the villeins, on condition that he would molest the city no more and would become co-burgher. But the nobles "adopted" by the city, and now residing

[1] Brentano fully understood the fatal effects of the struggle between the "old burghers" and the new-comers. Miaskowski, in his work on the village communities of Switzerland, has indicated the same for village communities.

within its walls, simply carried on the old war within the very precincts of the city. They disliked to submit to a tribunal of simple artisans and merchants, and fought their old feuds in the streets. Each city had now its Colonnas and Orsinis, its Overstolzes and Wises. Drawing large incomes from the estates they had still retained, they surrounded themselves with numerous clients and feudalized the customs and habits of the city itself. And when discontent began to be felt in the artisan classes of the town, they offered their sword and their followers to settle the differences by a free fight, instead of letting the discontent find out the channels which it did not fail to secure itself in olden times.

The greatest and the most fatal error of most cities was to base their wealth upon commerce and industry, to the neglect of agriculture. They thus repeated the error which had once been committed by the cities of antique Greece, and they fell through it into the same crimes.[1] The estrangement of so many cities from the land necessarily drew them into a policy hostile to the land, which became more and more evident in the times of Edward the Third,[2] the French Jacqueries, the Hussite wars, and the Peasant War in Germany. On the other hand, a commercial policy involved them in distant enterprises. Colonies were founded by the Italians in the south-east, by German

[1] The trade in slaves kidnapped in the East was never discontinued in the Italian republics till the fifteenth century. Feeble traces of it are found also in Germany and elsewhere. See Cibrario. *Della schiavitù e del servaggio*, 2 vols. Milan, 1868; Professor Luchitzkiy, "Slavery and Russian Slaves in Florence in the Fourteenth and Fifteenth Centuries," in *Izvestia* of the Kieff University, 1885.

[2] J. R. Green's *History of the English People*, London, 1878, i. 455.

cities in the east, by Slavonian cities in the far north-east. Mercenary armies began to be kept for colonial wars, and soon for local defence as well. Loans were contracted to such an extent as to totally demoralize the citizens; and internal contests grew worse and worse at each election, during which the colonial politics in the interest of a few families was at stake The division into rich and poor grew deeper, and in the sixteenth century, in each city, the royal authority found ready allies and support among the poor.

And there is yet another cause of the decay of communal institutions, which stands higher and lies deeper than all the above. The history of the mediæval cities offers one of the most striking illustrations of the power of *ideas* and *principles* upon the destinies of mankind, and of the quite opposed results which are obtained when a deep modification of leading ideas has taken place. Self-reliance and federalism, the sovereignty of each group, and the construction of the political body from the simple to the composite, were the leading ideas in the eleventh century. But since that time the conceptions had entirely changed. The students of Roman law and the prelates of the Church, closely bound together since the time of Innocent the Third, had succeeded in paralyzing the idea—the antique Greek idea—which presided at the foundation of the cities. For two or three hundred years they taught from the pulpit, the University chair, and the judges' bench, that salvation must be sought for in a strongly-centralized State, placed under a semi-divine authority;[1] that *one* man can and must be the saviour of society, and that in the name of public salvation he can com-

[1] See the theories expressed by the Bologna lawyers, already at the Congress of Roncaglia in 1158.

mit any violence: burn men and women at the stake, make them perish under indescribable tortures, plunge whole provinces into the most abject misery. Nor did they fail to give object lessons to this effect on a grand scale, and with an unheard-of cruelty, wherever the king's sword and the Church's fire, or both at once, could reach. By these teachings and examples, continually repeated and enforced upon public attention, the very minds of the citizens had been shaped into a new mould. They began to find no authority too extensive, no killing by degrees too cruel, once it was "for public safety." And, with this new direction of mind and this new belief in one man's power, the old federalist principle faded away, and the very creative genius of the masses died out. The Roman idea was victorious, and in such circumstances the centralized State had in the cities a ready prey.

Florence in the fifteenth century is typical of this change. Formerly a popular revolution was the signal of a new departure. Now, when the people, brought to despair, insurged, it had constructive ideas no more; no fresh idea came out of the movement. A thousand representatives were put into the Communal Council instead of 400; 100 men entered the *signoria* instead of 80. But a revolution of figures could be of no avail. The people's discontent was growing up, and new revolts followed. A saviour—the "tyran" —was appealed to; he massacred the rebels, but the disintegration of the communal body continued worse than ever. And when, after a new revolt, the people of Florence appealed to their most popular man, Gieronimo Savonarola, for advice, the monk's answer was:—"Oh, people mine, thou knowest that I cannot go into State affairs . . . . purify thy soul, and if in such a disposition of mind thou reformest thy city, then, people of Florence, thou shalt have inaugurated the reform in all Italy!" Carnival masks and vicious books were burned, a law of charity and another against usurers were passed—and the democracy of Florence remained where it was. The old spirit had gone. By too much trusting to government, they had ceased to trust to themselves; they were unable to open new issues. The State had only to step in and to crush down their last liberties.

And yet, the current of mutual aid and support did not die out in the masses, it continued to flow even after that defeat. It rose up again with a formidable force, in answer to the communist appeals of the first propagandists of the reform, and it continued to exist even after the masses, having failed to realize the life which they hoped to inaugurate under the inspiration of a reformed religion, fell under the dominions of an autocratic power. It flows still even now, and it seeks its way to find out a new expression which would not be the State, nor the mediæval city, nor the village community of the barbarians, nor the savage clan, but would proceed from all of them, and yet be superior to them in its wider and more deeply humane conceptions.

CHAPTER VII

MUTUAL AID AMONGST OURSELVES

Popular revolts at the beginning of the State-period.—Mutual Aid institutions of the present time.—The village community: its struggles for resisting its abolition by the State.—Habits derived from the village-community life, retained in our modern villages.—Switzerland, France, Germany, Russia.

THE mutual-aid tendency in man has so remote an origin, and is so deeply interwoven with all the past evolution of the human race, that it has been maintained by mankind up to the present time, notwithstanding all vicissitudes of history. It was chiefly evolved during periods of peace and prosperity; but when even the greatest calamities befell men—when whole countries were laid waste by wars, and whole populations were decimated by misery, or groaned under the yoke of tyranny—the same tendency continued to live in the villages and among the poorer classes in the towns; it still kept them together, and in the long run it reacted even upon those ruling, fighting, and devastating minorities which dismissed it as sentimental nonsense. And whenever mankind had to work out a new social organization, adapted to a new phasis of development, its constructive genius always drew the elements and the inspiration for the new departure from that same ever-living tendency. New economical and social institutions, in so far as they were a creation of the masses, new ethical systems, and new religions, all have originated from the same source, and the ethical progress of our race, viewed in its broad lines, appears as a gradual extension of the mutual-aid principles from the tribe to always larger and larger agglomerations, so as to finally embrace one day the whole of mankind, without respect to its divers creeds, languages, and races.

After having passed through the savage tribe, and next through the village community, the Europeans came to work out in mediæval times a new form of organization, which had the advantage of allowing great latitude for individual initiative, while it largely responded at the same time to man's need of mutual support. A federation of village communities, covered by a network of guilds and fraternities, was called into existence in the mediæval cities. The immense results achieved under this new form of union—in well-being for all, in industries, art, science, and commerce—were discussed at some length in two preceding chapters, and an attempt was also made to show why, towards the end of the fifteenth century, the mediæval republics—surrounded by domains of hostile feudal lords, unable to free the peasants from servitude, and gradually corrupted by ideas of Roman Cæsarism—were doomed to become a prey to the growing military States.

However, before submitting for three centuries to come, to the all-absorbing authority of the State, the masses of the people made a formidable attempt at reconstructing society on the old basis of mutual aid and support. It is well known by this time that the great movement of the reform was not a mere revolt against the abuses of the Catholic Church. It had its constructive ideal as well, and that ideal was life in free, brotherly communities. Those of the early writings and sermons of the period which found most response with the masses were imbued with ideas of the economical and social brotherhood of mankind. The "Twelve Articles" and similar professions of faith, which were circulated among the German and Swiss peasants and artisans, maintained not only every one's right to interpret the Bible according to his own understanding, but also included the demand of communal lands being restored to the village communities and feudal servitudes being abolished, and they always alluded to the "true" faith—a faith of brotherhood. At the same time scores of thousands of men and women joined the communist fraternities of Moravia, giving them all their fortune and living in numerous and prosperous settlements constructed upon the principles of communism.[1] Only wholesale massacres by the thousand could put a stop to this widely-spread popular movement, and it was by the sword, the fire, and the rack that the young States secured their first and decisive victory over the masses of the people.[2]

[1] A bulky literature, dealing with this formerly much-neglected subject, is now growing in Germany. Keller's works, *Ein Apostel der Wiedertäufer* and *Geschichte der Wiedertäufer*, Cornelius's *Geschichte des münsterischen Aufruhrs*, and Janssen's *Geschichte des deutschen Volkes* may be named as the leading sources. The first attempt at familiarizing English readers with the results of the wide researches made in Germany in this direction has been made in an excellent little work by Richard Heath—"Anabaptism from its Rise at Zwickau to its Fall at Münster, 1521–1536," London, 1895 (*Baptist Manuals*, vol. i.)—where the leading features of the movement are well indicated, and full bibliographical information is given. Also K. Kautsky's "Communism in Central Europe in the Time of the Reformation," London, 1897.

[2] Few of our contemporaries realize both the extent of this movement and the means by which it was suppressed. But those who wrote immediately after the great peasant war estimated at from 100,000 to 150,000 men the number of peasants slaughtered after

For the next three centuries the States, both on the Continent and in these islands, systematically weeded out all institutions in which the mutual-aid tendency had formerly found its expression. The village communities were bereft of their folkmotes, their courts and independent administration; their lands were confiscated. The guilds were spoliated of their possessions and liberties, and placed under the control, the fancy, and the bribery of the State's official. The cities were divested of their sovereignty, and the very springs of their inner life—the folkmote, the elected justices and administration, the sovereign parish and the sovereign guild—were annihilated; the State's functionary took possession of every link of what formerly was an organic whole. Under that fatal policy and the wars it engendered, whole regions, once populous and wealthy, were laid bare; rich cities became insignificant boroughs; the very roads which connected them with other cities became impracticable. Industry, art, and knowledge fell into decay. Political education, science, and law were rendered subservient to the idea of State centralization. It was taught in the Universities and from the pulpit that the institutions in which men formerly used to embody their needs of mutual support could not be tolerated in a properly organized State; that the State alone could represent the bonds of union between its subjects; that federalism and "particularism" were the enemies of progress, and the State was the only proper initiator of further development. By the end of the last century the kings on the Continent, the Parliament in these

their defeat in Germany. See Zimmermann's *Allgemeine Geschichte des grossen Bauernkrieges*. For the measures taken to suppress the movement in the Netherlands see Richard Heath's *Anabaptism*.

isles, and the revolutionary Convention in France, although they were at war with each other, agreed in asserting that no separate unions between citizens must exist within the State; that hard labour and death were the only suitable punishments to workers who dared to enter into "coalitions." "No state within the State!" The State alone, and the State's Church, must take care of matters of general interest, while the subjects must represent loose aggregations of individuals, connected by no particular bonds, bound to appeal to the Government each time that they feel a common need. Up to the middle of this century this was the theory and practice in Europe. Even commercial and industrial societies were looked at with suspicion. As to the workers, their unions were treated as unlawful almost within our own lifetime in this country and within the last twenty years on the Continent. The whole system of our State education was such that up to the present time, even in this country, a notable portion of society would treat as a revolutionary measure the concession of such rights as every one, freeman or serf, exercised five hundred years ago in the village folkmote, the guild, the parish, and the city.

The absorption of all social functions by the State necessarily favoured the development of an unbridled, narrow-minded individualism. In proportion as the obligations towards the State grew in numbers the citizens were evidently relieved from their obligations towards each other. In the guild—and in mediæval times every man belonged to some guild or fraternity—two "brothers" were bound to watch in turns a brother who had fallen ill; it would be sufficient now to give one's neighbour the address of the next paupers' hospital. In barbarian society, to assist at a fight between two men, arisen from a quarrel, and not to prevent it from taking a fatal issue, meant to be oneself treated as a murderer; but under the theory of the all-protecting State the bystander need not intrude: it is the policeman's business to interfere, or not. And while in a savage land, among the Hottentots, it would be scandalous to eat without having loudly called out thrice whether there is not somebody wanting to share the food, all that a respectable citizen has to do now is to pay the poor tax and to let the starving starve. The result is, that the theory which maintains that men can, and must, seek their own happiness in a disregard of other people's wants is now triumphant all round—in law, in science, in religion. It is the religion of the day, and to doubt of its efficacy is to be a dangerous Utopian. Science loudly proclaims that the struggle of each against all is the leading principle of nature, and of human societies as well. To that struggle Biology ascribes the progressive evolution of the animal world. History takes the same line of argument; and political economists, in their naïve ignorance, trace all progress of modern industry and machinery to the "wonderful" effects of the same principle. The very religion of the pulpit is a religion of individualism, slightly mitigated by more or less charitable relations to one's neighbours, chiefly on Sundays. "Practical" men and theorists, men of science and religious preachers, lawyers and politicians, all agree upon one thing—that individualism may be more or less softened in its harshest effects by charity, but that it is the only secure basis for the maintenance of society and its ulterior progress.

It seems, therefore, hopeless to look for mutual-aid institutions and practices in modern society. What could remain of them? And yet, as soon as we try to ascertain how the millions of human beings live, and begin to study their everyday relations, we are struck with the immense part which the mutual-aid and mutual-support principles play even now-a-days in human life. Although the destruction of mutual-aid institutions has been going on in practice and theory, for full three or four hundred years, hundreds of millions of men continue to live under such institutions; they piously maintain them and endeavour to reconstitute them where they have ceased to exist. In our mutual relations every one of us has his moments of revolt against the fashionable individualistic creed of the day, and actions in which men are guided by their mutual-aid inclinations constitute so great a part of our daily intercourse that if a stop to such actions could be put all further ethical progress would be stopped at once. Human society itself could not be maintained for even so much as the lifetime of one single generation. These facts, mostly neglected by sociologists and yet of the first importance for the life and further elevation of mankind, we are now going to analyze, beginning with the standing institutions of mutual support, and passing next to those acts of mutual aid which have their origin in personal or social sympathies.

When we cast a broad glance on the present constitution of European society we are struck at once with the fact that, although so much has been done to get rid of the village community, this form of union continues to exist to the extent we shall presently see, and that many attempts are now made either to reconstitute it in some shape or another or to find some substitute for it. The current theory as regards the village community is, that in Western Europe it has died out by a natural death, because the communal possession of the soil was found inconsistent with the modern requirements of agriculture. But the truth is that nowhere did the village community disappear of its own accord; everywhere, on the contrary, it took the ruling classes several centuries of persistent but not always successful efforts to abolish it and to confiscate the communal lands.

In France, the village communities began to be deprived of their independence, and their lands began to be plundered, as early as the sixteenth century. However, it was only in the next century, when the mass of the peasants was brought, by exactions and wars, to the state of subjection and misery which is vividly depicted by all historians, that the plundering of their lands became easy and attained scandalous proportions. "Every one has taken of them according to his powers . . . Imaginary debts have been claimed, in order to seize upon their lands;" so we read in an edict promulgated by Louis the Fourteenth in 1667.[1] Of course the State's remedy for such evils was to render the communes still more subservient to the State, and to plunder them itself. In fact, two years later all money revenue of the communes was confiscated by the King. As to the appropriation of communal lands, it grew worse and worse, and in the next century the nobles and the clergy had already taken possession of immense tracts of land—one-half of the cultivated area, according to certain estimates

[1] "Chacun s'en est accommodé selon sa bienséance . . . on les a partagés . . . pour dépouiller les communes, on s'est servi de dettes simulées" (Edict of Louis the Fourteenth, of 1667, quoted by several authors. Eight years before that date the communes had been taken under State management).

—mostly to let it go out of culture.[1] But the peasants still maintained their communal institutions, and until the year 1787 the village folkmotes, composed of all householders, used to come together in the shadow of the bell-tower or a tree, to allot and re-allot what they had retained of their fields, to assess the taxes, and to elect their executive, just as the Russian *mir* does at the present time. This is what Babeau's researches have proved to demonstration.[2]

The Government found, however, the folkmotes "too noisy," too disobedient, and in 1787, elected councils, composed of a mayor and three to six syndics, chosen from among the wealthier peasants, were introduced instead. Two years later the Revolutionary Assemblée Constituante, which was on this point at one with the old *régime*, fully confirmed this law (on the 14th of December, 1789), and the *bourgeois du village* had now their turn for the plunder of communal lands, which continued all through the Revolutionary period. Only on the 16th of August, 1792, the Convention, under the pressure of the peasants' insurrections, decided to return the enclosed lands to the communes;[3] but it ordered at the same time that they should be divided in equal parts among the wealthier peasants only—a measure which pro-

[1] "On a great landlord's estate, even if he has millions of revenue, you are sure to find the land uncultivated" (Arthur Young). "One-fourth part of the soil went out of culture;" "for the last hundred years the land has returned to a savage state;" "the formerly flourishing Sologne is now a big marsh;" and so on (Théron de Montaugé, quoted by Taine in *Origines de la France Contemporaine*, tome i. p. 441).

[2] A. Babeau, *Le Village sous l'Ancien Régime*, 3e édition. Paris, 1892.

[3] In Eastern France the law only confirmed what the peasants had already done themselves; in other parts of France it usually remained a dead letter.

voked new insurrections and was abrogated next year, in 1793, when the order came to divide the communal lands among all commoners, rich and poor alike, "active" and "inactive."

These two laws, however, ran so much against the conceptions of the peasants that they were not obeyed, and wherever the peasants had retaken possession of part of their lands they kept them undivided. But then came the long years of wars, and the communal lands were simply confiscated by the State (in 1794) as a mortgage for State loans, put up for sale, and plundered as such; then returned again to the communes and confiscated again (in 1813); and only in 1816 what remained of them, *i. e.* about 15,000,000 acres of the least productive land, was restored to the village communities.[1] Still this was not yet the end of the troubles of the communes. Every new *régime* saw in the communal lands a means for gratifying its supporters, and three laws (the first in 1837 and the last under Napoleon the Third) were passed to induce the village communities to divide their estates. Three

[1] After the triumph of the middle-class reaction the communal lands were declared (August 24, 1794) the States domains, and, together with the lands confiscated from the nobility, were put up for sale, and pilfered by the *bandes noires* of the small *bourgeoisie.* True that a stop to this pilfering was put next year (law of 2 Prairial, An V), and the preceding law was abrogated; but then the village communities were simply abolished, and cantonal councils were introduced instead. Only seven years later (9 Prairial, An XII), *i. e.* in 1801, the village communities were reintroduced, but not until after having been deprived of all their rights, the mayor and syndics being nominated by the Government in the 36,000 communes of France! This system was maintained till after the revolution of 1830, when elected communal councils were reintroduced under the law of 1787. As to the communal lands, they were again seized upon by the State in 1813, plundered as such, and only partly restored to the communes in 1816. See the classical collection of French laws, by Dalloz, *Répertoire de Jurisprudence;* also the works of Doniol, Dareste, Bonnemère, Babeau, and many others.

times these laws had to be repealed, in consequence of the opposition they met with in the villages; but something was snapped up each time, and Napoleon the Third, under the pretext of encouraging perfected methods of agriculture, granted large estates out of the communal lands to some of his favourites.

As to the autonomy of the village communities, what could be retained of it after so many blows? The mayor and the syndics were simply looked upon as unpaid functionaries of the State machinery. Even now, under the Third Republic, very little can be done in a village community without the huge State machinery, up to the *préfet* and the ministries, being set in motion. It is hardly credible, and yet it is true, that when, for instance, a peasant intends to pay in money his share in the repair of a communal road, instead of himself breaking the necessary amount of stones, no fewer than twelve different functionaries of the State must give their approval, and an aggregate of *fifty-two* different acts must be performed by them, and exchanged between them, before the peasant is permitted to pay that money to the communal council. All the remainder bears the same character.[1]

What took place in France took place everywhere in Western and Middle Europe. Even the chief dates of the great assaults upon the peasant lands are the same. For England the only difference is that the spoliation was accomplished by separate acts rather than by general sweeping measures—with less haste but more thoroughly than in France. The

[1] This procedure is so absurd that one would not believe it possible if the fifty-two different acts were not enumerated in full by a quite authoritative writer in the *Journal des Economistes* (1893, April, p. 94), and several similar examples were not given by the same author.

seizure of the communal lands by the lords also began in the fifteenth century, after the defeat of the peasant insurrection of 1380—as seen from Rossus's *Historia* and from a statute of Henry the Seventh, in which these seizures are spoken of under the heading of "enormitees and myschefes as be hurtfull . . . to the common wele."[1] Later on the Great Inquest, under Henry the Eighth, was begun, as is known, in order to put a stop to the enclosure of communal lands, but it ended in a sanction of what had been done.[2] The communal lands continued to be preyed upon, and the peasants were driven from the land. But it was especially since the middle of the eighteenth century that, in England as everywhere else, it became part of a systematic policy to simply weed out all traces of communal ownership; and the wonder is not that it has disappeared, but that it could be maintained, even in England, so as to be "generally prevalent so late as the grandfathers of this generation."[3] The very object of the Enclosure Acts, as shown by Mr. Seebohm, was to remove this system,[4] and it was so

[1] Dr. Ochenkowski, *Englands wirthschaftliche Entwickelung im Ausgange des Mittelalters* (Jena, 1879), pp. 35 *seq.*, where the whole question is discussed with full knowledge of the texts.

[2] Nasse, *Ueber die mittelalterliche Feldgemeinschaft und die Einhegungen des XVI. Jahrhunderts in England* (Bonn, 1869), pp. 4, 5; Vinogradov, *Villainage in England* (Oxford, 1892).

[3] F. Seebohm, *The English Village Community*, 3rd edition, 1884, pp. 13–15.

[4] "An examination into the details of an Enclosure Act will make clear the point that the system as above described [communal ownership] is the system which it was the object of the Enclosure Act to remove" (Seebohm, *l. c.* p. 13). And further on, "They were generally drawn in the same form, commencing with the recital that the open and common fields lie dispersed in small pieces, intermixed with each other and inconveniently situated; that divers persons own parts of them, and are entitled to rights of common on them . . . and that it is desired that they may be divided and

well removed by the nearly four thousand Acts passed between 1760 and 1844 that only faint traces of it remain now. The land of the village communities was taken by the lords, and the appropriation was sanctioned by Parliament in each separate case.

In Germany, in Austria, in Belgium the village community was also destroyed by the State. Instances of commoners themselves dividing their lands were rare,[1] while everywhere the States coerced them to enforce the division, or simply favoured the private appropriation of their lands. The last blow to communal ownership in Middle Europe also dates from the middle of the eighteenth century. In Austria sheer force was used by the Government, in 1768, to compel the communes to divide their lands—a special commission being nominated two years later for that purpose. In Prussia Frederick the Second, in several of his ordinances (in 1752, 1763, 1765, and 1769), recommended to the *Justizcollegien* to enforce the division. In Silesia a special resolution was issued to serve that aim in 1771. The same took place in Belgium, and, as the communes did not obey, a law was issued in 1847 empowering the Government to buy communal meadows in order to sell them in retail, and to make a forced sale of the communal land when there was a would-be buyer for it.[2]

---

enclosed, a specific share being let out and allowed to each owner" (p. 14). Porter's list contained 3867 such Acts, of which the greatest numbers fall upon the decades of 1770–1780 and 1800–1820, as in France.

[1] In Switzerland we see a number of communes, ruined by wars, which have sold part of their lands, and now endeavour to buy them back.

[2] A. Buchenberger, "Agrarwesen und Agrarpolitik," in A. Wagner's *Handbuch der politischen Oekonomie*, 1892, Band i. pp. 280 *seq.*

In short, to speak of the natural death of the village communities in virtue of economical laws is as grim a joke as to speak of the natural death of soldiers slaughtered on a battlefield. The fact was simply this: The village communities had lived for over a thousand years; and where and when the peasants were not ruined by wars and exactions they steadily improved their methods of culture. But as the value of land was increasing, in consequence of the growth of industries, and the nobility had acquired, under the State organization, a power which it never had had under the feudal system, it took possession of the best parts of the communal lands, and did its best to destroy the communal institutions.

However, the village-community institutions so well respond to the needs and conceptions of the tillers of the soil that, in spite of all, Europe is up to this date covered with *living* survivals of the village communities, and European country life is permeated with customs and habits dating from the community period. Even in England, notwithstanding all the drastic measures taken against the old order of things, it prevailed as late as the beginning of the nineteenth century. Mr. Gomme—one of the very few English scholars who have paid attention to the subject—shows in his work that many traces of the communal possession of the soil are found in Scotland, "runrig" tenancy having been maintained in Forfarshire up to 1813, while in certain villages of Inverness the custom was, up to 1801, to plough the land for the whole community, without leaving any boundaries, and to allot it after the ploughing was done. In Kilmorie the allotment and re-allotment of the fields was in full vigour "till the last twenty-five years," and the Crofters' Commission found it still in vigour in certain islands.[1] In Ireland the system prevailed up to the great famine; and as to England, Marshall's works, which passed unnoticed until Nasse and Sir Henry Maine drew attention to them, leave no doubt as to the village-community system having been widely spread, in nearly all English counties, at the beginning of the nineteenth century.[2] No more than twenty years ago Sir Henry Maine was "greatly surprised at the number of instances of abnormal property rights, necessarily implying the former existence of collective ownership and joint cultivation," which a comparatively brief inquiry brought under his notice.[3] And, communal institutions having persisted so late as that, a great number of mutual-aid habits and customs would undoubtedly be discovered in English villages if the writers of this country only paid attention to village life.[4]

[1] G. L. Gomme, "The Village Community, with special reference to its Origin and Forms of Survival in Great Britain" (*Contemporary Science Series*), London, 1890, pp. 141–143; also his *Primitive Folkmoots* (London, 1880), pp. 98 *seq.*

[2] "In almost all parts of the country, in the Midland and Eastern counties particularly, but also in the west—in Wiltshire, for example —in the south, as in Surrey, in the north, as in Yorkshire,—there are extensive open and common fields. Out of 316 parishes of Northamptonshire 89 are in this condition; more than 100 in Oxfordshire; about 50,000 arces in Warwickshire; in Berkshire half the county; more than half of Wiltshire; in Huntingdonshire out of a total area of 240,000 acres 130,000 were commonable meadows, commons, and fields" (Marshall, quoted in Sir Henry Maine's *Village Communities in the East and West*, New York edition, 1876, pp. 88, 89).

[3] *Ibid.* p. 88; also Fifth Lecture. The wide extension of "commons" in Surrey, even now, is well known.

[4] In quite a number of books dealing with English country life which I have consulted I have found charming descriptions of country scenery and the like, but almost nothing about the daily life and customs of the labourers.

As to the Continent, we find the communal institutions fully alive in many parts of France, Switzerland, Germany, Italy, the Scandinavian lands, and Spain, to say nothing of Eastern Europe; the village life in these countries is permeated with communal habits and customs; and almost every year the Continental literature is enriched by serious works dealing with this and connected subjects. I must, therefore, limit my illustrations to the most typical instances. Switzerland is undoubtedly one of them. Not only the five republics of Uri, Schwytz, Appenzell, Glarus, and Unterwalden hold their lands as undivided estates, and are governed by their popular folkmotes, but in all other cantons too the village communities remain in possession of a wide self-government, and own large parts of the Federal territory.[1] Two-thirds of all the Alpine meadows and two-thirds of all the forests of Switzerland are until now communal land; and a considerable number of fields, orchards, vineyards, peat bogs, quarries, and so on, are owned in common. In the Vaud, where all the householders continue to take part in the deliberations of their elected communal councils, the communal spirit is especially alive. Towards the end of the winter all the young men of each village go to stay a few days in the woods, to fell timber and to bring it down the steep slopes tobogganing way, the timber and the

[1] In Switzerland the peasants in the open land also fell under the dominion of lords, and large parts of their estates were appropriated by the lords in the sixteenth and seventeenth centuries. (See, for instance, Dr. A. Miaskowski, in Schmoller's *Forschungen*, Bd. ii. 1879, pp. 12 *seq.*) But the peasant war in Switzerland did not end in such a crushing defeat of the peasants as it did in other countries, and a great deal of the communal rights and lands was retained. The self-government of the communes is, in fact, the very foundation of the Swiss liberties.

fuel wood being divided among all households or sold for their benefit. These excursions are real *fêtes* of manly labour. On the banks of Lake Leman part of the work required to keep up the terraces of the vineyards is still done in common; and in the spring, when the thermometer threatens to fall below zero before sunrise, the watchman wakes up all householders, who light fires of straw and dung and protect their vine-trees from the frost by an artificial cloud. In nearly all cantons the village communities possess so-called *Bürgernutzen*—that is, they hold in common a number of cows, in order to supply each family with butter; or they keep communal fields or vineyards, of which the produce is divided between the burghers; or they rent their land for the benefit of the community.[1]

It may be taken as a rule that where the communes have retained a wide sphere of functions, so as to be living parts of the national organism, and where they have not been reduced to sheer misery, they never fail to take good care of their lands. Accordingly the communal estates in Switzerland strikingly contrast with the miserable state of "commons" in this country. The communal forests in the Vaud and the Valais are admirably managed, in conformity with the rules of modern forestry. Elsewhere the "strips" of communal fields, which change owners under the system of re-allotment, are very well manured, especially as there is no lack of meadows and cattle. The high-level meadows are well kept as a rule, and the rural roads are excellent.[2] And when we admire the Swiss

[1] Miaskowski, in Schmoller's *Forschungen*, Bd. ii. 1879, p. 15.

[2] See on this subject a series of works, summed up in one of the excellent and suggestive chapters (not yet translated into English) which K. Bücher has added to the German translation of Laveleye's *Primitive Ownership*. Also Meitzen, "Das Agrar- und Forst-Wesen,

châlet, the mountain road, the peasants' cattle, the terraces of vineyards, or the school-house in Switzerland, we must keep in mind that without the timber for the *châlet* being taken from the communal woods and the stone from the communal quarries, without the cows being kept on the communal meadows, and the roads being made and the school-houses built by communal work, there would be little to admire.

It hardly need be said that a great number of mutual-aid habits and customs continue to persist in the Swiss villages. The evening gatherings for shelling walnuts, which take place in turns in each household; the evening parties for sewing the dowry of the girl who is going to marry; the calling of "aids" for building the houses and taking in the crops, as well as for all sorts of work which may be required by one of the commoners; the custom of exchanging children from one canton to the other, in order to make them learn two languages, French and German; and so on—all these are quite habitual;[1] while, on the other side, divers modern requirements are met in the same spirit. Thus in Glarus most of the Alpine meadows have been sold during a time of calamity; but the communes still continue to buy field land, and after the newly-bought fields have been left in the possession of separate commoners for ten, twenty, or thirty years, as the case might be, they return to the common stock, which is re-allotted

die Allmenden und die Landgemeinden der Deutschen Schweiz," in *Jahrbuch für Staatswissenschaft*, 1880, iv. (analysis of Miaskowsky's works); O'Brien, "Notes in a Swiss village," in *Macmillan's Magazine*, October 1885.

[1] The wedding gifts, which often substantially contribute in this country to the comfort of the young households, are evidently a remainder of the communal habits.

according to the needs of all. A great number of small associations are formed to produce some of the necessaries for life—bread, cheese, and wine—by common work, be it only on a limited scale; and agricultural co-operation altogether spreads in Switzerland with the greatest ease. Associations formed between ten to thirty peasants, who buy meadows and fields in common, and cultivate them as co-owners, are of common occurrence; while dairy associations for the sale of milk, butter, and cheese are organized everywhere. In fact, Switzerland was the birthplace of that form of co-operation. It offers, moreover, an immense field for the study of all sorts of small and large societies, formed for the satisfaction of all sorts of modern wants. In certain parts of Switzerland one finds in almost every village a number of associations—for protection from fire, for boating, for maintaining the quays on the shores of a lake, for the supply of water, and so on; and the country is covered with societies of archers, sharpshooters, topographers, footpath explorers, and the like, originated from modern militarism.

Switzerland is, however, by no means an exception in Europe, because the same institutions and habits are found in the villages of France, of Italy, of Germany, of Denmark, and so on. We have just seen what has been done by the rulers of France in order to destroy the village community and to get hold of its lands; but notwithstanding all that one-tenth part of the whole territory available for culture, *i.e.* 13,500,000 acres, including one-half of all the natural meadows and nearly a fifth part of all the forests of the country, remain in communal possession. The woods supply the communers with fuel, and the timber wood is cut, mostly by communal work, with all desirable regularity; the grazing lands are free for the commoners' cattle; and what remains of communal fields is allotted and re-allotted in certain parts of France—namely, in the Ardennes—in the usual way.[1]

These additional sources of supply, which aid the poorer peasants to pass through a year of bad crops without parting with their small plots of land and without running into irredeemable debts, have certainly their importance for both the agricultural labourers and the nearly three millions of small peasant proprietors. It is even doubtful whether small peasant proprietorship could be maintained without these additional resources. But the ethical importance of the communal possessions, small as they are, is still greater than their economical value. They maintain in village life a nucleus of customs and habits of mutual aid which undoubtedly acts as a mighty check upon the development of reckless individualism and greediness, which small land-ownership is only too prone to develop. Mutual aid in all possible circumstances of village life is part of the routine life in all parts of the country. Everywhere we meet, under different names, with the *charroi*, i. e. the free aid of the neighbours for taking in a crop, for vintage, or for building a house; everywhere we find the same evening gatherings as have just been mentioned in Switzerland; and everywhere the commoners associate for all sorts of work. Such habits are mentioned by

[1] The communes own, 4,554,100 acres of woods out of 24,813,000 in the whole territory, and 6,936,300 acres of natural meadows out of 11,394,000 acres in France. The remaining 2,000,000 acres are fields, orchards, and so on.

nearly all those who have written upon French village life. But it will perhaps be better to give in this place some abstracts from letters which I have just received from a friend of mine whom I have asked to communicate to me his observations on this subject. They come from an aged man who for years has been the mayor of his commune in South France (in Ariège); the facts he mentions are known to him from long years of personal observation, and they have the advantage of coming from one neighbourhood instead of being skimmed from a large area. Some of them may seem trifling, but as a whole they depict quite a little world of village life.

"In several communes in our neighbourhood," my friend writes, "the old custom of *l'emprount* is in vigour. When many hands are required in a *métairie* for rapidly making some work—dig out potatoes or mow the grass—the youth of the neighbourhood is convoked; young men and girls come in numbers, make it gaily and for nothing; and in the evening, after a gay meal, they dance.

"In the same communes, when a girl is going to marry, the girls of the neighbourhood come to aid in sewing the dowry. In several communes the women still continue to spin a good deal. When the winding off has to be done in a family it is done in one evening—all friends being convoked for that work. In many communes of the Ariège and other parts of the south-west the shelling of the Indian corn-sheaves is also done by all the neighbours. They are treated with chestnuts and wine, and the young people dance after the work has been done. The same custom is practised for making nut oil and crushing hemp. In the commune of L. the same is done for bringing in the corn crops. These days of hard work become *fête* days, as the owner stakes his honour on serving a good meal. No remuneration is given; all do it for each other.[1]

"In the commune of S. the common grazing-land is every year increased, so that nearly the whole of the land of the commune is now kept in common. The shepherds are elected by all owners of the cattle, including women. The bulls are communal.

"In the commune of M. the forty to fifty small sheep flocks of the commoners are brought together and divided into three or four flocks before being sent to the higher meadows. Each owner goes for a week to serve as shepherd.

"In the hamlet of C. a threshing machine has been bought in common by several households; the fifteen to twenty persons required to serve the machine being supplied by all the families. Three other threshing machines have been bought and are rented out by their owners, but the work is performed by outside helpers, invited in the usual way.

"In our commune of R. we had to raise the wall of the cemetery. Half of the money which was required for buying lime and for the wages of the skilled workers was supplied by the county council, and the other half by subscription. As to the work of carrying sand and water, making mortar, and serving the masons, it was done entirely by volunteers [just as in the Kabyle *djemmâa*]. The rural roads were repaired in the same way, by volunteer days of work given by the commoners. Other communes have built in

[1] In Caucasia they even do better among the Georgians. As the meal costs, and a poor man cannot afford to give it, a sheep is bought by those same neighbours who come to aid in the work.

the same way their fountains. The wine-press and other smaller appliances are frequently kept by the commune."

Two residents of the same neighbourhood, questioned by my friend, add the following:—

"At O. a few years ago there was no mill. The commune has built one, levying a tax upon the commoners. As to the miller, they decided, in order to avoid frauds and partiality, that he should be paid two francs for each bread-eater, and the corn be ground free.

"At St. G. few peasants are insured against fire. When a conflagration has taken place—so it was lately—all give something to the family which has suffered from it—a chaldron, a bed-cloth, a chair, and so on—and a modest household is thus reconstituted. All the neighbours aid to build the house, and in the meantime the family is lodged free by the neighbours."

Such habits of mutual support—of which many more examples could be given—undoubtedly account for the easiness with which the French peasants associate for using, in turn, the plough with its team of horses, the wine-press, and the threshing machine, when they are kept in the village by one of them only, as well as for the performance of all sorts of rural work in common. Canals were maintained, forests were cleared, trees were planted, and marshes were drained by the village communities from time immemorial; and the same continues still. Quite lately, in *La Borne* of Lozère barren hills were turned into rich gardens by communal work. "The soil was brought on men's backs; terraces were made and planted with chestnut trees, peach trees, and orchards, and water was brought for irrigation in canals two or three miles long." Just now they have dug a new canal, eleven miles in length.[1]

To the same spirit is also due the remarkable success lately obtained by the *syndicats agricoles*, or peasants' and farmers' associations. It was not until 1884 that associations of more than nineteen persons were permitted in France, and I need not say that when this "dangerous experiment" was ventured upon—so it was styled in the Chambers—all due "precautions" which functionaries can invent were taken. Notwithstanding all that, France begins to be covered with syndicates. At the outset they were only formed for buying manures and seeds, falsification having attained colossal proportions in these two branches;[2] but gradually they extended their functions in various directions, including the sale of agricultural produce and permanent improvements of the land. In South France the ravages of the phylloxera have called into existence a great number of wine-growers' associations. Ten to thirty growers form a syndicate, buy a steam-engine for pumping water, and make the necessary arrangements for inundating their vineyards in turn.[3] New associations for protecting the land

[1] Alfred Baudrillart, in H. Baudrillart's *Les Populations Rurales de la France*, 3rd series (Paris, 1893), p. 479.

[2] The *Journal des Économistes* (August 1892, May and August 1893) has lately given some of the results of analyses made at the agricultural laboratories at Ghent and at Paris. The extent of falsification is simply incredible; so also the devices of the "honest traders." In certain seeds of grass there was 32 per cent. of grains of sand, coloured so as to deceive even an experienced eye; other samples contained from 52 to 22 per cent. only of pure seed, the remainder being weeds. Seeds of vetch contained 11 per cent. of a poisonous grass (*nielle*); a flour for cattle-fattening contained 36 per cent. of sulphates; and so on *ad infinitum*.

[3] A. Baudrillart, *l. c.* p. 309. Originally one grower would undertake to supply water, and several others would agree to make use of it. "What especially characterises such associations," A. Bau-

from inundations, for irrigation purposes, and for maintaining canals are continually formed, and the unanimity of all peasants of a neighbourhood, which is required by law, is no obstacle. Elsewhere we have the *fruitières*, or dairy associations, in some of which all butter and cheese is divided in equal parts, irrespective of the yield of each cow. In the Ariège we find an association of eight separate communes for the common culture of their lands, which they have put together; syndicates for free medical aid have been formed in 172 communes out of 337 in the same department; associations of consumers arise in connection with the syndicates; and so on.[1] "Quite a revolution is going on in our villages," Alfred Baudrillart writes, "through these associations, which take in each region their own special characters."

Very much the same must be said of Germany. Wherever the peasants could resist the plunder of their lands, they have retained them in communal ownership, which largely prevails in Württemberg, Baden, Hohenzollern, and in the Hessian province of Starkenberg.[2] The communal forests are kept, as a

---

drillart remarks, "is that no sort of written agreement is concluded. All is arranged in words. There was, however, not one single case of difficulties having arisen between the parties."

[1] A. Baudrillart, *l. c.* pp. 300, 341, etc. M. Terssac, president of the St. Gironnais syndicate (Ariège), wrote to my friend in substance as follows:—"For the exhibition of Toulouse our association has grouped the owners of cattle which seemed to us worth exhibiting. The society undertook to pay one-half of the travelling and exhibition expenses; one-fourth was paid by each owner, and the remaining fourth by those exhibitors who had got prizes. The result was that many took part in the exhibition who never would have done it otherwise. Those who got the highest awards (350 francs) have contributed 10 per cent. of their prizes, while those who have got no prize have only spent 6 to 7 francs each."

[2] In Württemberg 1,629 communes out of 1,910 have communal property. They owned in 1863 over 1,000,000 acres of land. In

rule, in an excellent state, and in thousands of communes timber and fuel wood are divided every year among all inhabitants; even the old custom of the *Lesholztag* is widely spread: at the ringing of the village bell all go to the forest to take as much fuel wood as they can carry.[1] In Westphalia one finds communes in which all the land is cultivated as one common estate, in accordance with all requirements of modern agronomy. As to the old communal customs and habits, they are in vigour in most parts of Germany. The calling in of *aids*, which are real *fêtes* of labour, is known to be quite habitual in Westphalia, Hesse, and Nassau. In well-timbered regions the timber for a new house is usually taken from the communal forest, and all the neighbours join in building the house. Even in the suburbs of Frankfort it is a regular custom among the gardeners that in case of one of them being ill all come on Sunday to cultivate his garden.[2]

In Germany, as in France, as soon as the rulers of the people repealed their laws against the peasant associations—that was only in 1884–1888—these unions began to develop with a wonderful rapidity, notwithstanding all legal obstacles which were put in

---

Baden 1,256 communes out of 1,582 have communal land; in 1884–1888 they held 121,500 acres of fields in communal culture, and 675,000 acres of forests, *i. e.* 46 per cent. of the total area under woods. In Saxony 39 per cent. of the total area is in communal ownership (Schmoller's *Jahrbuch*, 1886, p. 359). In Hohenzollern nearly two-thirds of all meadow land, and in Hohenzollern-Hechingen 41 per cent. of all landed property, are owned by the village communities (Buchenberger, *Agrarwesen*, vol. i. p. 300).

[1] See K. Bücher, who, in a special chapter added to Laveleye's *Ureigenthum*, has collected all information relative to the village community in Germany.

[2] K. Bücher, *ibid.* pp. 89, 90.

their way.[1] "It is a fact," Buchenberger says, "that in *thousands* of village communities, in which no sort of chemical manure or rational fodder was ever known, both have become of everyday use, to a quite unforeseen extent, owing to these associations" (vol. ii. p. 507). All sorts of labour-saving implements and agricultural machinery, and better breeds of cattle, are bought through the associations, and various arrangements for improving the quality of the produce begin to be introduced. Unions for the sale of agricultural produce are also formed, as well as for permanent improvements of the land.[2]

From the point of view of social economics all these efforts of the peasants certainly are of little importance. They cannot substantially, and still less permanently, alleviate the misery to which the tillers of the soil are doomed all over Europe. But from the ethical point of view, which we are now considering, their importance cannot be overrated. They prove that even under the system of reckless individualism which now prevails the agricultural masses piously maintain their mutual-support inheritance; and as soon as the States relax the iron laws by means of which they have broken all bonds between men, these bonds are at once reconstituted, notwithstanding the difficulties, political, economical, and social, which are many, and

[1] For this legislation and the numerous obstacles which were put in the way, in the shape of red-tapeism and supervision, see Buchenberger's *Agrarwesen und Agrarpolitik*, Bd. ii. pp. 342–363, and p. 506, note

[2] Buchenberger, *l. c.* Bd. ii. p. 510. The General Union of Agricultural Co-operation comprises an aggregate of 1,679 societies. In Silesia an aggregate of 32,000 acres of land has been lately drained by 73 associations; 454,800 acres in Prussia by 516 associations; in Bavaria there are 1,715 drainage and irrigation unions.

in such forms as best answer to the modern requirements of production. They indicate in which direction and in which form further progress must be expected.

I might easily multiply such illustrations, taking them from Italy, Spain, Denmark, and so on, and pointing out some interesting features which are proper to each of these countries.[1] The Slavonian populations of Austria and the Balkan peninsula, among whom the "compound family," or "undivided household," is found in existence, ought also to be mentioned.[2] But I hasten to pass on to Russia, where the same mutual-support tendency takes certain new and unforeseen forms. Moreover, in dealing with the village community in Russia we have the advantage of possessing an immense mass of materials, collected during the colossal house-to-house inquest which was lately made by several *zemstvos* (county councils), and which embraces a population of nearly 20,000,000 peasants in different parts of the country.[3]

Two important conclusions may be drawn from the bulk of evidence collected by the Russian inquests. In Middle Russia, where fully one-third of the peasants have been brought to utter ruin (by heavy taxation, small allotments of unproductive land, rack rents, and

[1] See Appendix XII.

[2] For the Balkan peninsula see Laveleye's *Propriété Primitive*.

[3] The facts concerning the village community, contained in nearly a hundred volumes (out of 450) of these inquests, have been classified and summed up in an excellent Russian work by "V. V.," *The Peasant Community* (*Krestianskaya Obschina*), St. Petersburg, 1892, which, apart from its theoretical value, is a rich compendium of data relative to this subject. The above inquests have also given origin to an immense literature, in which the modern village-community question for the first time emerges from the domain of generalities and is put on the solid basis of reliable and sufficiently detailed facts.

very severe tax-collecting after total failures of crops), there was, during the first five-and-twenty years after the emancipation of the serfs, a decided tendency towards the constitution of individual property in land within the village communities. Many impoverished "horseless" peasants abandoned their allotments, and this land often became the property of those richer peasants, who borrow additional incomes from trade, or of outside traders, who buy land chiefly for exacting rack rents from the peasants. It must also be added that a flaw in the land-redemption law of 1861 offered great facilities for buying peasants' lands at a very small expense,[1] and that the State officials mostly used their weighty influence in favour of individual as against communal ownership. However, for the last twenty years a strong wind of opposition to the individual appropriation of the land blows again through the Middle Russian villages, and strenuous efforts are being made by the bulk of those peasants who stand between the rich and the very poor to uphold the village community. As to the fertile steppes of the South, which are now the most populous and the richest part of European Russia, they were mostly colonized, during the present century, under the system of individual ownership or occupation, sanctioned in that form by the State. But since improved methods of agriculture with the aid of machinery have been introduced in the region, the peasant owners have

[1] The redemption had to be paid by annuities for forty-nine years. As years went, and the greatest part of it was paid, it became easier and easier to redeem the smaller remaining part of it, and, as each allotment could be redeemed individually, advantage was taken of this disposition by traders, who bought land for half its value from the ruined peasants. A law was consequently passed to put a stop to such sales.

gradually begun themselves to transform their individual ownership into communal possession, and one finds now, in that granary of Russia, a very great number of spontaneously formed village communities of recent origin.[1]

The Crimea and the part of the mainland which lies to the north of it (the province of Taurida), for which we have detailed data, offer an excellent illustration of that movement. This territory began to be colonized, after its annexation in 1783, by Great, Little, and White Russians—Cossacks, freemen, and runaway serfs—who came individually or in small groups from all corners of Russia. They took first to cattle-breeding, and when they began later on to till the soil, each one tilled as much as he could afford to. But when—immigration continuing, and perfected ploughs being introduced—land stood in great demand, bitter disputes arose among the settlers. They lasted for years, until these men, previously tied by no mutual bonds, gradually came to the idea that an end must be put to disputes by introducing village-community ownership. They passed decisions to the effect that the land which they owned individually should henceforward be their common property, and they began to allot and to re-allot it in accordance with the usual village-community rules. The movement gradually took a great extension, and on a small territory, the Taurida statisticians found 161 villages in which communal ownership had been introduced by the peasant proprietors themselves, chiefly in the years 1855–1885, *in*

[1] Mr. V. V., in his *Peasant Community*, has grouped together all facts relative to this movement. About the rapid agricultural development of South Russia and the spread of machinery English readers will find information in the Consular Reports (Odessa, Taganrog).

lieu of individual ownership. Quite a variety of village-community types has been freely worked out in this way by the settlers.[1] What adds to the interest of this transformation is that it took place, not only among the Great Russians, who are used to village-community life, but also among Little Russians, who have long since forgotten it under Polish rule, among Greeks and Bulgarians, and even among Germans, who have long since worked out in their prosperous and half-industrial Volga colonies their own type of village community.[2] It is evident that the Mussulman Tartars of Taurida hold their land under the Mussulman customary law, which is limited personal occupation; but even with them the European village community has been introduced in a few cases. As to other nationalities in Taurida, individual ownership has been abolished in six Esthonian, two Greek, two Bulgarian, one Czech, and one German village.

This movement is characteristic for the whole of the fertile steppe region of the south. But separate instances of it are also found in Little Russia. Thus in a number of villages of the province of Chernigov the peasants were formerly individual owners of their plots; they had separate legal documents for their plots and used to rent and to sell their land at will. But in the fifties of the nineteenth century a movement began among them in favour of communal possession, the chief

[1] In some instances they proceeded with great caution. In one village they began by putting together all meadow land, but only a small portion of the fields (about five acres per soul) was rendered communal; the remainder continued to be owned individually. Later on, in 1862–1864, the system was extended, but only in 1884 was communal possession introduced in full.—V. V.'s *Peasant Community*, pp. 1–14.

[2] On the Mennonite village community see A. Klaus, *Our Colonies* (*Nashi Kolonii*), St. Petersburg, 1869.

argument being the growing number of pauper families. The initiative of the reform was taken in one village, and the others followed suit, the last case on record dating from 1882. Of course there were struggles between the poor, who usually claim for communal possession, and the rich, who usually prefer individual ownership; and the struggles often lasted for years. In certain places the unanimity required then by the law being impossible to obtain, the village divided into two villages, one under individual ownership and the other under communal possession; and so they remained until the two coalesced into one community, or else they remained divided still. As to Middle Russia, it is a fact that in many villages which were drifting towards individual ownership there began since 1880 a mass movement in favour of re-establishing the village community. Even peasant proprietors who had lived for years under the individualist system returned *en masse* to the communal institutions. Thus, there is a considerable number of ex-serfs who have received one-fourth part only of the regulation allotments, but they have received them free of redemption and in individual ownership. There was in 1890 a wide-spread movement among them (in Kursk, Ryazan, Tambov, Orel, etc.) towards putting their allotments together and introducing the village community. The "free agriculturists" (*volnyie khlebopashtsy*), who were liberated from serfdom under the law of 1803, and had *bought* their allotments—each family separately—are now nearly all under the village-community system, which they have introduced themselves. All these movements are of recent origin, and non-Russians too join them. Thus the Bulgares in the district of Tiraspol, after having remained for sixty years under the personal-property system, introduced the village community in the years 1876–1882. The German Mennonites of Berdyansk fought in 1890 for introducing the village community, and the small peasant proprietors (*Kleinwirthschaftliche*) among the German Baptists were agitating in their villages in the same direction. One instance more: In the province of Samara the Russian government created in the forties, by way of experiment, 103 villages on the system of individual ownership. Each household received a splendid property of 105 acres. In 1890, out of the 103 villages the peasants in 72 had already notified the desire of introducing the village community. I take all these facts from the excellent work of V. V., who simply gives, in a classified form, the facts recorded in the above-mentioned house-to-house inquest.

This movement in favour of communal possession runs badly against the current economical theories, according to which intensive culture is incompatible with the village community. But the most charitable thing that can be said of these theories is that they have never been submitted to the test of experiment: they belong to the domain of political metaphysics. The facts which we have before us show, on the contrary, that wherever the Russian peasants, owing to a concurrence of favourable circumstances, are less miserable than they are on the average, and wherever they find men of knowledge and initiative among their neighbours, the village community becomes the very means for introducing various improvements in agriculture and village life altogether. Here, as elsewhere, mutual aid is a better leader to progress than the war of each against all, as may be seen from the following facts.

Under Nicholas the First's rule many Crown officials and serf-owners used to compel the peasants to introduce the communal culture of small plots of the village lands, in order to refill the communal storehouses after loans of grain had been granted to the poorest commoners. Such cultures, connected in the peasants' minds with the worst reminiscences of serfdom, were abandoned as soon as serfdom was abolished; but now the peasants begin to reintroduce them on their own account. In one district (Ostrogozhsk, in Kursk) the initiative of one person was sufficient to call them to life in four-fifths of all the villages. The same is met with in several other localities. On a given day the commoners come out, the richer ones with a plough or a cart and the poorer ones single-handed, and no attempt is made to discriminate one's share in the work. The crop is afterwards used for loans to the poorer commoners, mostly free grants, or for the orphans and widows, or for the village church, or for the school, or for repaying a communal debt.[1]

That all sorts of work which enters, so to say, in the routine of village life (repair of roads and bridges, dams, drainage, supply of water for irrigation, cutting of wood, planting of trees, etc.) are made by whole communes, and that land is rented and meadows are mown by whole communes—the work being accomplished by old and young, men and women, in the way described by Tolstoi—is only what one may expect

[1] Such communal cultures are known to exist in 159 villages out of 195 in the Ostrogozhsk district; in 150 out of 187 in Slavyano-serbsk; in 107 village communities in Alexandrovsk, 93 in Nikolayevsk, 35 in Elisabethgrad. In a German colony the communal culture is made for repaying a communal debt. All join in the work, although the debt was contracted by 94 householders out of 155.

from people living under the village-community system.[1] They are of everyday occurrence all over the country. But the village community is also by no means averse to modern agricultural improvements, when it can stand the expense, and when knowledge, hitherto kept for the rich only, finds its way into the peasant's house.

It has just been said that perfected ploughs rapidly spread in South Russia, and in many cases the village communities were instrumental in spreading their use. A plough was bought by the community, experimented upon on a portion of the communal land, and the necessary improvements were indicated to the makers, whom the communes often aided in starting the manufacture of cheap ploughs as a village industry. In the district of Moscow, where 1,560 ploughs were lately bought by the peasants during five years, the impulse came from those communes which rented lands as a body for the special purpose of improved culture.

In the north-east (Vyatka) small associations of peasants, who travel with their winnowing machines (manufactured as a village industry in one of the iron districts), have spread the use of such machines in the neighbouring governments. The very wide spread of threshing machines in Samara, Saratov, and Kherson is due to the peasant associations, which can afford to buy a costly engine, while the individual peasant cannot. And while we read in nearly all economical treatises that the village community was doomed to disappear when the three-fields system had to be substituted by the rotation of crops system, we see in Russia many village communities taking the initiative of introducing the rotation of crops. Before accepting

[1] Lists of such works which came under the notice of the *zemstvo* statisticians will be found in V. V.'s *Peasant Community*, pp. 459-600.

it the peasants usually set apart a portion of the communal fields for an experiment in artificial meadows, and the commune buys the seeds.[1] If the experiment proves successful they find no difficulty whatever in re-dividing their fields, so as to suit the five or six fields system.

This system is now in use in *hundreds* of villages of Moscow, Tver, Smolensk, Vyatka, and Pskov.[2] And where land can be spared the communities give also a portion of their domain to allotments for fruit-growing. Finally, the sudden extension lately taken in Russia by the little model farms, orchards, kitchen gardens, and silkworm-culture grounds—which are started at the village school-houses, under the conduct of the school-master, or of a village volunteer—is also due to the support they found with the village communities.

Moreover, such permanent improvements as drainage and irrigation are of frequent occurrence. For instance, in three districts of the province of Moscow—industrial to a great extent—drainage works have been accomplished within the last ten years on a large

[1] In the government of Moscow the experiment was usually made on the field which was reserved for the above-mentioned communal culture.

[2] Several instances of such and similar improvements were given in the *Official Messenger*, 1894, Nos. 256–258. Associations between "horseless" peasants begin to appear also in South Russia. Another extremely interesting fact is the sudden development in Southern West Siberia of very numerous co-operative creameries for making butter. Hundreds of them spread in Tobolsk and Tomsk, without any one knowing wherefrom the initiative of the movement came. It came from the Danish co-operators, who used to export their own butter of higher quality, and to buy butter of a lower quality for their own use in Siberia. After a several years' intercourse, they introduced creameries there. Now, a great export trade has grown out of their endeavours.

scale in no less than 180 to 200 different villages—the commoners working themselves with the spade. At another extremity of Russia, in the dry Steppes of Novouzen, over a thousand dams for ponds were built and several hundreds of deep wells were sunk by the communes; while in a wealthy German colony of the south-east the commoners worked, men and women alike, for five weeks in succession, to erect a dam, two miles long, for irrigation purposes. What could isolated men do in that struggle against the dry climate? What could they obtain through individual effort when South Russia was struck with the marmot plague, and all people living on the land, rich and poor, commoners and individualists, had to work with their hands in order to conjure the plague? To call in the policeman would have been of no use; to associate was the only possible remedy.

And now, after having said so much about mutual aid and support which are practised by the tillers of the soil in "civilized" countries, I see that I might fill an octavo volume with illustrations taken from the life of the hundreds of millions of men who also live under the tutorship of more or less centralized States, but are out of touch with modern civilization and modern ideas. I might describe the inner life of a Turkish village and its network of admirable mutual-aid customs and habits. On turning over my leaflets covered with illustrations from peasant life in Caucasia, I come across touching facts of mutual support. I trace the same customs in the Arab *djemmâa* and the Afghan *purra*, in the villages of Persia, India, and Java, in the undivided family of the Chinese, in the encampments of the semi-nomads of Central Asia and the nomads of the far North. On consulting notes taken at random in the literature of Africa, I find them replete with similar facts—of aids convoked to take in the crops, of houses built by all inhabitants of the village—sometimes to repair the havoc done by civilized filibusters—of people aiding each other in case of accident, protecting the traveller, and so on. And when I peruse such works as Post's compendium of African customary law I understand why, notwithstanding all tyranny, oppression, robberies and raids, tribal wars, glutton kings, deceiving witches and priests, slave-hunters, and the like, these populations have not gone astray in the woods; why they have maintained a certain civilization, and have remained men, instead of dropping to the level of straggling families of decaying orang-outans. The fact is, that the slave-hunters, the ivory robbers, the fighting kings, the Matabele and the Madagascar "heroes" pass away, leaving their traces marked with blood and fire; but the nucleus of mutual-aid institutions, habits, and customs, grown up in the tribe and the village community, remains; and it keeps men united in societies, open to the progress of civilization, and ready to receive it when the day comes that they shall receive civilization instead of bullets.

The same applies to our civilized world. The natural and social calamities pass away. Whole populations are periodically reduced to misery or starvation; the very springs of life are crushed out of millions of men, reduced to city pauperism; the understanding and the feelings of the millions are vitiated by teachings worked out in the interest of the few. All this is certainly a part of our existence. But the nucleus of mutual-support institutions, habits, and customs remains alive with the millions ; it keeps them together ; and they prefer to cling to their customs, beliefs, and traditions rather than to accept the teachings of a war of each against all, which are offered to them under the title of science, but are no science at all.

CHAPTER VIII

MUTUAL AID AMONGST OURSELVES (*continued*)

Labour-unions grown after the destruction of the guilds by the State.—Their struggles.—Mutual Aid in strikes.—Co-operation.—Free associations for various purposes.—Self-sacrifice.—Countless societies for combined action under all possible aspects.—Mutual Aid in slum-life.—Personal aid.

WHEN we examine the every-day life of the rural populations of Europe, we find that, notwithstanding all that has been done in modern States for the destruction of the village community, the life of the peasants remains honeycombed with habits and customs of mutual aid and support; that important vestiges of the communal possession of the soil are still retained; and that, as soon as the legal obstacles to rural association were lately removed, a network of free unions for all sorts of economical purposes rapidly spread among the peasants—the tendency of this young movement being to reconstitute some sort of union similar to the village community of old. Such being the conclusions arrived at in the preceding chapter, we have now to consider, what institutions for mutual support can be found at the present time amongst the industrial populations.

For the last three hundred years, the conditions for the growth of such institutions have been as unfavourable in the towns as they have been in the villages.

It is well known, indeed, that when the mediæval cities were subdued in the sixteenth century by growing military States, all institutions which kept the artisans, the masters, and the merchants together in the guilds and the cities were violently destroyed. The self-government and the self-jurisdiction of both the guild and the city were abolished; the oath of allegiance between guild-brothers became an act of felony towards the State; the properties of the guilds were confiscated in the same way as the lands of the village communities; and the inner and technical organization of each trade was taken in hand by the State. Laws, gradually growing in severity, were passed to prevent artisans from combining in any way. For a time, some shadows of the old guilds were tolerated: merchants' guilds were allowed to exist under the condition of freely granting subsidies to the kings, and some artisan guilds were kept in existence as organs of administration. Some of them still drag on their meaningless existence. But what formerly was the vital force of mediæval life and industry has long since disappeared under the crushing weight of the centralized State.

In Great Britain, which may be taken as the best illustration of the industrial policy of the modern States, we see the Parliament beginning the destruction of the guilds as early as the fifteenth century; but it was especially in the next century that decisive measures were taken. Henry the Eighth not only ruined the organization of the guilds, but also confiscated their properties, with even less excuse and manners, as Toulmin Smith wrote, than he had produced for confiscating the estates of the monasteries.[1]

[1] Toulmin Smith, *English Guilds*, London, 1870, Introd. p. xliii.

Edward the Sixth completed his work,[1] and already in the second part of the sixteenth century we find the Parliament settling all the disputes between craftsmen and merchants, which formerly were settled in each city separately. The Parliament and the king not only legislated in all such contests, but, keeping in view the interests of the Crown in the exports, they soon began to determine the number of apprentices in each trade and minutely to regulate the very technics of each fabrication—the weights of the stuffs, the number of threads in the yard of cloth, and the like. With little success, it must be said; because contests and technical difficulties which were arranged for centuries in succession by agreement between closely-interdependent guilds and federated cities lay entirely beyond the powers of the centralized State. The continual interference of its officials paralyzed the trades, bringing most of them to a complete decay; and the last century economists, when they rose against the State regulation of industries, only ventilated a widely-felt discontent. The abolition of that interference by the French Revolution was greeted as an act of liberation, and the example of France was soon followed elsewhere.

With the regulation of wages the State had no better success. In the mediæval cities, when the distinction between masters and apprentices or journeymen became more and more apparent in the

[1] The Act of Edward the Sixth—the first of his reign—ordered to hand over to the Crown "all fraternities, brotherhoods, and guilds being within the realm of England and Wales and other of the king's dominions; and all manors, lands, tenements, and other hereditaments belonging to them or any of them" (*English Guilds*, Introd. p. xliii). See also Ockenkowski's *Englands wirtschaftliche Entwickelung im Ausgange des Mittelalters*, Jena, 1879, chaps. ii.–v.

fifteenth century, unions of apprentices (*Gesellenverbände*), occasionally assuming an international character, were opposed to the unions of masters and merchants. Now it was the State which undertook to settle their griefs, and under the Elizabethan Statute of 1563 the Justices of Peace had to settle the wages, so as to guarantee a "convenient" livelihood to journeymen and apprentices. The Justices, however, proved helpless to conciliate the conflicting interests, and still less to compel the masters to obey their decisions. The law gradually became a dead letter, and was repealed by the end of the eighteenth century. But while the State thus abandoned the function of regulating wages, it continued severely to prohibit all combinations which were entered upon by journeymen and workers in order to raise their wages, or to keep them at a certain level. All through the eighteenth century it legislated against the workers' unions, and in 1799 it finally prohibited all sorts of combinations, under the menace of severe punishments. In fact, the British Parliament only followed in this case the example of the French Revolutionary Convention, which had issued a draconic law against coalitions of workers—coalitions between a number of citizens being considered as attempts against the sovereignty of the State, which was supposed equally to protect all its subjects. The work of destruction of the mediæval unions was thus completed. Both in the town and in the village the State reigned over loose aggregations of individuals, and was ready to prevent by the most stringent measures the reconstitution of any sort of separate unions among them. These were, then, the conditions under which the mutual-aid tendency had to make its way in the nineteenth century.

Need it be said that no such measures could destroy that tendency? Throughout the eighteenth century, the workers' unions were continually reconstituted.[1] Nor were they stopped by the cruel prosecutions which took place under the laws of 1797 and 1799. Every flaw in supervision, every delay of the masters in denouncing the unions was taken advantage of. Under the cover of friendly societies, burial clubs, or secret brotherhoods, the unions spread in the textile industries, among the Sheffield cutlers, the miners, and vigorous federal organizations were formed to support the branches during strikes and prosecutions.[2]

The repeal of the Combination Laws in 1825 gave a new impulse to the movement. Unions and national federations were formed in all trades;[3] and when Robert Owen started his Grand National Consolidated Trades' Union, it mustered half a million members in a few months. True that this period of relative liberty did not last long. Prosecution began anew in the thirties, and the well-known ferocious condemnations of 1832–1844 followed. The Grand National Union was disbanded, and all over the country, both the private employers and the Government in its own workshops began to compel the workers to resign all

[1] See Sidney and Beatrice Webb, *History of Trade-Unionism*, London, 1894, pp. 21–38.

[2] See in Sidney Webb's work the associations which existed at that time. The London artisans are supposed to have never been better organized than in 1810–20.

[3] The National Association for the Protection of Labour included about 150 separate unions, which paid high levies, and had a membership of about 100,000. The Builders' Union and the Miners' Unions also were big organizations (Webb, *l.c.* p. 107).

connection with unions, and to sign "the Document" to that effect. Unionists were prosecuted wholesale under the Master and Servant Act—workers being summarily arrested and condemned upon a mere complaint of misbehaviour lodged by the master.[1] Strikes were suppressed in an autocratic way, and the most astounding condemnations took place for merely having announced a strike or acted as a delegate in it—to say nothing of the military suppression of strike riots, nor of the condemnations which followed the frequent outbursts of acts of violence. To practise mutual support under such circumstances was anything but an easy task. And yet, notwithstanding all obstacles, of which our own generation hardly can have an idea, the revival of the unions began again in 1841, and the amalgamation of the workers has been steadily continued since. After a long fight, which lasted for over a hundred years, the right of combining together was conquered, and at the present time nearly one-fourth part of the regularly-employed workers, *i. e.* about 1,500,000, belong to trade unions.[2]

As to the other European States, sufficient to say that up to a very recent date, all sorts of unions were

[1] I follow in this Mr. Webb's work, which is replete with documents to confirm his statements.

[2] Great changes have taken place since the forties in the attitude of the richer classes towards the unions. However, even in the sixties, the employers made a formidable concerted attempt to crush them by locking out whole populations. Up to 1869 the simple agreement to strike, and the announcement of a strike by placards, to say nothing of picketing, were often punished as intimidation. Only in 1875 the Master and Servant Act was repealed, peaceful picketing was permitted, and "violence and intimidation" during strikes fell into the domain of common law. Yet, even during the dock-labourers' strike in 1887, relief money had to be spent for fighting before the Courts for the right of picketing, while the prosecutions of the last few years menace once more to render the conquered rights illusory.

prosecuted as conspiracies; and that nevertheless they exist everywhere, even though they must often take the form of secret societies; while the extension and the force of labour organizations, and especially of the Knights of Labour, in the United States and in Belgium, have been sufficiently illustrated by strikes in the nineties. It must, however, be borne in mind that, prosecution apart, the mere fact of belonging to a labour union implies considerable sacrifices in money, in time, and in unpaid work, and continually implies the risk of losing employment for the mere fact of being a unionist.[1] There is, moreover, the strike, which a unionist has continually to face; and the grim reality of a strike is, that the limited credit of a worker's family at the baker's and the pawnbroker's is soon exhausted, the strike-pay goes not far even for food, and hunger is soon written on the children's faces. For one who lives in close contact with workers, a protracted strike is the most heartrending sight; while what a strike meant forty years ago in this country, and still means in all but the wealthiest parts of the continent, can easily be conceived. Continually, even now, strikes will end with the total ruin and the forced emigration of whole populations, while the shooting down of strikers on the slightest provocation, or even without any provocation,[2] is quite habitual still on the continent.

[1] A weekly contribution of *6d.* out of an *18s.* wage, or of *1s.* out of *25s.*, means much more than *9l.* out of a *300l.* income: it is mostly taken upon food; and the levy is soon doubled when a strike is declared in a brother union. The graphic description of trade-union life, by a skilled craftsman, published by Mr. and Mrs. Webb (pp. 431 *seq.*), gives an excellent idea of the amount of work required from a unionist.

[2] See the debates upon the strikes of Falkenau in Austria before the Austrian Reichstag on the 10th of May, 1894, in which debates

And yet, every year there are thousands of strikes and lock-outs in Europe and America—the most severe and protracted contests being, as a rule, the so-called "sympathy strikes," which are entered upon to support locked-out comrades or to maintain the rights of the unions. And while a portion of the Press is prone to explain strikes by "intimidation," those who have lived among strikers speak with admiration of the mutual aid and support which are constantly practised by them. Every one has heard of the colossal amount of work which was done by volunteer workers for organizing relief during the London dock-labourers' strike; of the miners who, after having themselves been idle for many weeks, paid a levy of four shillings a week to the strike fund when they resumed work; of the miner widow who, during the Yorkshire labour war of 1894, brought her husband's life-savings to the strike-fund; of the last loaf of bread being always shared with neighbours; of the Radstock miners, favoured with larger kitchen-gardens, who invited four hundred Bristol miners to take their share of cabbage and potatoes, and so on. All newspaper correspondents, during the great strike of miners in Yorkshire in 1894, knew heaps of such facts, although not all of them could report such "irrelevant" matters to their respective papers.[1]

Unionism is not, however, the only form in which the worker's need of mutual support finds its expression. There are, besides, the political associations, whose activity many workers consider as more con-

the fact is fully recognized by the Ministry and the owner of the colliery. Also the English Press of that time.

[1] Many such facts will be found in the *Daily Chronicle* and partly the *Daily News* for October and November 1894.

ducive to general welfare than the trade-unions, limited as they are now in their purposes. Of course the mere fact of belonging to a political body cannot be taken as a manifestation of the mutual-aid tendency. We all know that politics are the field in which the purely egotistic elements of society enter into the most entangled combinations with altruistic aspirations. But every experienced politician knows that all great political movements were fought upon large and often distant issues, and that those of them were the strongest which provoked most disinterested enthusiasm. All great historical movements have had this character, and for our own generation Socialism stands in that case. "Paid agitators" is, no doubt, the favourite refrain of those who know nothing about it. The truth, however, is that—to speak only of what I know personally—if I had kept a diary for the last twenty-four years and inscribed in it all the devotion and self-sacrifice which I came across in the Socialist movement, the reader of such a diary would have had the word "heroism" constantly on his lips. But the men I would have spoken of were not heroes; they were average men, inspired by a grand idea. Every Socialist newspaper—and there are hundreds of them in Europe alone—has the same history of years of sacrifice without any hope of reward, and, in the overwhelming majority of cases, even without any personal ambition. I have seen families living without knowing what would be their food to-morrow, the husband boycotted all round in his little town for his part in the paper, and the wife supporting the family by sewing, and such a situation lasting for years, until the family would retire, without a word of reproach, simply saying: "Continue; we can hold on no more!" I have seen men, dying from consumption, and knowing it, and yet knocking about in snow and fog to prepare meetings, speaking at meetings within a few weeks from death, and only then retiring to the hospital with the words: "Now, friends, I am done; the doctors say I have but a few weeks to live. Tell the comrades that I shall be happy if they come to see me." I have seen facts which would be described as "idealization" if I told them in this place; and the very names of these men, hardly known outside a narrow circle of friends, will soon be forgotten when the friends, too, have passed away. In fact, I don't know myself which most to admire, the unbounded devotion of these few, or the sum total of petty acts of devotion of the great number. Every quire of a penny paper sold, every meeting, every hundred votes which are won at a Socialist election, represent an amount of energy and sacrifices of which no outsider has the faintest idea. And what is now done by Socialists has been done in every popular and advanced party, political and religious, in the past. All past progress has been promoted by like men and by a like devotion.

Co-operation, especially in Britain, is often described as "joint-stock individualism"; and such as it is now, it undoubtedly tends to breed a co-operative egotism, not only towards the community at large, but also among the co-operators themselves. It is, nevertheless, certain that at its origin the movement had an essentially mutual-aid character. Even now, its most ardent promoters are persuaded that co-operation leads mankind to a higher harmonic stage of economical relations, and it is not possible to stay in some of the strongholds of co-operation in the North without realizing that the great number of the rank and file hold the same opinion. Most of them would lose interest in the movement if that faith were gone; and it must be owned that within the last few years broader ideals of general welfare and of the producers' solidarity have begun to be current among the co-operators. There is undoubtedly now a tendency towards establishing better relations between the owners of the co-operative workshops and the workers.

The importance of co-operation in this country, in Holland and in Denmark is well known; while in Germany, and especially on the Rhine, the co-operative societies are already an important factor of industrial life.[1] It is, however, Russia which offers perhaps the best field for the study of co-operation under an infinite variety of aspects. In Russia, it is a natural growth, an inheritance from the middle ages; and while a formally established co-operative society would have to cope with many legal difficulties and official suspicion, the informal co-operation—the *artél*—makes the very substance of Russian peasant life. The history of "the making of Russia," and of the colonization of Siberia, is a history of the hunting and trading *artéls* or guilds, followed by village communities, and at the present time we find the *artél* everywhere; among each group of ten to fifty peasants who come from the same village to work at a factory, in all the building trades, among fishermen and hunters, among convicts on their way to and in Siberia, among railway porters, Exchange messengers, Customs House labourers, everywhere in the village

[1] The 31,473 productive and consumers' associations on the Middle Rhine showed, about 1890, a yearly expenditure of 18,437,500*l.*; 3,675,000*l.* were granted during the year in loans.

industries, which give occupation to 7,000,000 men—from top to bottom of the working world, permanent and temporary, for production and consumption under all possible aspects. Until now, many of the fishing-grounds on the tributaries of the Caspian Sea are held by immense *artéls*, the Ural river belonging to the whole of the Ural Cossacks, who allot and re-allot the fishing-grounds—perhaps the richest in the world—among the villages, without any interference of the authorities. Fishing is always made by *artéls* in the Ural, the Volga, and all the lakes of Northern Russia. Besides these permanent organizations, there are the simply countless temporary *artéls*, constituted for each special purpose. When ten or twenty peasants come from some locality to a big town, to work as weavers, carpenters, masons, boat-builders, and so on, they always constitute an *artél*. They hire rooms, hire a cook (very often the wife of one of them acts in this capacity), elect an elder, and take their meals in common, each one paying his share for food and lodging to the *artél*. A party of convicts on its way to Siberia always does the same, and its elected elder is the officially-recognized intermediary between the convicts and the military chief of the party. In the hard-labour prisons they have the same organization. The railway porters, the messengers at the Exchange, the workers at the Custom House, the town messengers in the capitals, who are collectively responsible for each member, enjoy such a reputation that any amount of money or bank-notes is trusted to the *artél*-member by the merchants. In the building trades, *artéls* of from 10 to 200 members are formed; and the serious builders and railway contractors always prefer to deal with an *artél* than with separately-hired workers. The last attempts of the Ministry of War to deal directly with productive *artéls*, formed *ad hoc* in the domestic trades, and to give them orders for boots and all sorts of brass and iron goods, are described as most satisfactory; while the renting of a Crown iron work (*Votkinsk*) to an *artél* of workers, which took place seven or eight years ago, has been a decided success.

We can thus see in Russia how the old mediæval institution, having not been interfered with by the State (in its informal manifestations), has fully survived until now, and takes the greatest variety of forms in accordance with the requirements of modern industry and commerce. As to the Balkan peninsula, the Turkish Empire and Caucasia, the old guilds are maintained there in full. The *esnafs* of Servia have fully preserved their mediæval character; they include both masters and journeymen, regulate the trades, and are institutions for mutual support in labour and sickness;[1] while the *amkari* of Caucasia, and especially at Tiflis, add to these functions a considerable influence in municipal life.[2]

In connection with co-operation, I ought perhaps to mention also the friendly societies, the unities of oddfellows, the village and town clubs organized for meeting the doctors' bills, the dress and burial clubs, the small clubs very common among factory girls, to which they contribute a few pence every week, and afterwards draw by lot the sum of one pound, which can at least be used for some substantial purchase, and many others.

[1] British Consular Report, April 1889.

[2] A capital research on this subject has been published in Russian in the *Zapiski* (*Memoirs*) of the Caucasian Geographical Society, vol. vi. 2, Tiflis, 1891, by C. Egiazaroff.

A not inconsiderable amount of sociable or jovial spirit is alive in all such societies and clubs, even though the "credit and debit" of each member are closely watched over. But there are so many associations based on the readiness to sacrifice time, health, and life if required, that we can produce numbers of illustrations of the best forms of mutual support.

The Lifeboat Association in this country, and similar institutions on the Continent, must be mentioned in the first place. The former has now over three hundred boats along the coasts of these isles, and it would have twice as many were it not for the poverty of the fishermen, who cannot afford to buy lifeboats. The crews consist, however, of volunteers, whose readiness to sacrifice their lives for the rescue of absolute strangers to them is put every year to a severe test; every winter the loss of several of the bravest among them stands on record. And if we ask these men what moves them to risk their lives, even when there is no reasonable chance of success, their answer is something on the following lines. A fearful snowstorm, blowing across the Channel, raged on the flat, sandy coast of a tiny village in Kent, and a small smack, laden with oranges, stranded on the sands near by. In these shallow waters only a flat-bottomed lifeboat of a simplified type can be kept, and to launch it during such a storm was to face an almost certain disaster. And yet the men went out, fought for hours against the wind, and the boat capsized twice. One man was drowned, the others were cast ashore. One of these last, a refined coastguard, was found next morning, badly bruised and half frozen in the snow. I asked him, how they came to make that desperate attempt? "I don't know myself," was his reply. "*There* was the wreck; all the people from the village stood on the beach, and all said it would be foolish to go out; we never should work through the surf. We saw five or six men clinging to the mast, making desperate signals. We all felt that something must be done, but what could we do? One hour passed, two hours, and we all stood there. We all felt most uncomfortable. Then, all of a sudden, through the storm, it seemed to us as if we heard their cries—they had a boy with them. We could not stand that any longer. All at once we said, "We must go!" The women said so too; they would have treated us as cowards if we had not gone, although next day they said we had been fools to go. As one man, we rushed to the boat, and went. The boat capsized, but we took hold of it. The worst was to see poor —— drowning by the side of the boat, and we could do nothing to save him. Then came a fearful wave, the boat capsized again, and we were cast ashore. The men were still rescued by the D. boat, ours was caught miles away. I was found next morning in the snow."

The same feeling moved also the miners of the Rhonda Valley, when they worked for the rescue of their comrades from the inundated mine. They had pierced through thirty-two yards of coal in order to reach their entombed comrades; but when only three yards more remained to be pierced, fire-damp enveloped them. The lamps went out, and the rescue-men retired. To work in such conditions was to risk being blown up at every moment. But the raps of the entombed miners were still heard, the men were still alive and appealed for help, and several miners volunteered to work at any risk; and as they went down the mine, their wives had only silent tears to follow them—not one word to stop them.

There is the gist of human psychology. Unless men are maddened in the battlefield, they "cannot stand it" to hear appeals for help, and not to respond to them. The hero goes; and what the hero does, *all* feel that they ought to have done as well. The sophisms of the brain cannot resist the mutual-aid feeling, because this feeling has been nurtured by thousands of years of human social life and hundreds of thousands of years of pre-human life in societies.

"But what about those men who were drowned in the Serpentine in the presence of a crowd, out of which no one moved for their rescue?" it may be asked. "What about the child which fell into the Regent's Park Canal—also in the presence of a holiday crowd—and was only saved through the presence of mind of a maid who let out a Newfoundland dog to the rescue?" The answer is plain enough. Man is a result of both his inherited instincts and his education. Among the miners and the seamen, their common occupations and their every-day contact with one another create a feeling of solidarity, while the surrounding dangers maintain courage and pluck. In the cities, on the contrary, the absence of common interest nurtures indifference, while courage and pluck, which seldom find their opportunities, disappear, or take another direction. Moreover, the tradition of the hero of the mine and the sea lives in the miners' and fishermen's villages, adorned with a poetical halo. But what are the traditions of a motley London crowd? The only tradition they might have in common ought to be created by literature, but a literature which would correspond to the village epics hardly exists. The clergy are so anxious to prove that all that comes from human nature is sin, and that all good in man has a supernatural origin, that they mostly ignore the facts which cannot be produced as an example of higher inspiration or grace, coming from above. And as to the lay-writers, their attention is chiefly directed towards one sort of heroism, the heroism which promotes the idea of the State. Therefore, they admire the Roman hero, or the soldier in the battle, while they pass by the fisherman's heroism, hardly paying attention to it. The poet and the painter might, of course, be taken by the beauty of the human heart in itself; but both seldom know the life of the poorer classes, and while they can sing or paint the Roman or the military hero in conventional surroundings, they can neither sing nor paint impressively the hero who acts in those modest surroundings which they ignore. If they venture to do so, they produce a mere piece of rhetoric.[1]

[1] Escape from a French prison is extremely difficult; nevertheless a prisoner escaped from one of the French prisons in 1884 or 1885. He even managed to conceal himself during the whole day, although the alarm was given and the peasants in the neighbourhood were on the look-out for him. Next morning found him concealed in a ditch, close by a small village. Perhaps he intended to steal some food, or some clothes in order to take off his prison uniform. As he was lying in the ditch a fire broke out in the village. He saw a woman running out of one of the burning houses, and heard her desperate appeals to rescue a child in the upper storey of the burning house. No one moved to do so. Then the escaped prisoner dashed out of his retreat, made his way through the fire, and, with a scalded face and burning clothes, brought the child safe out of the fire, and handed it to its mother. Of course he was arrested on the spot by the village *gendarme*, who now made his appearance. He was taken back to the prison. The fact was reported in all French papers, but none of them bestirred itself to obtain his release. If he had shielded a warder from a comrade's blow, he would have been made a hero of. But his act was simply humane, it did not promote the

The countless societies, clubs, and alliances, for the enjoyment of life, for study and research, for education, and so on, which have lately grown up in such numbers that it would require many years to simply tabulate them, are another manifestation of the same ever-working tendency for association and mutual support. Some of them, like the broods of young birds of different species which come together in the autumn, are entirely given to share in common the joys of life. Every village in this country, in Switzerland, Germany, and so on, has its cricket, football, tennis, nine-pins, pigeon, musical or singing clubs. Other societies are much more numerous, and some of them, like the Cyclists' Alliance, have suddenly taken a formidable development. Although the members of this alliance have nothing in common but the love of cycling, there is already among them a sort of freemasonry for mutual help, especially in the remote nooks and corners which are not flooded by cyclists; they look upon the "C.A.C."—the Cyclists' Alliance Club—in a village as a sort of home; and at the yearly Cyclists' Camp many a standing friendship has been established. The *Kegelbrüder*, the Brothers of the Nine Pins, in Germany, are a similar association; so also the Gymnasts' Societies (300,000 members in Germany), the informal brotherhood of paddlers in France, the yacht clubs, and so on. Such associations certainly do not alter the economical stratification of society, but, especially in the small towns, they contribute to smooth social distinctions, and as they all tend to

State's ideal; he himself did not attribute it to a sudden inspiration of divine grace; and that was enough to let the man fall into oblivion. Perhaps, six or twelve months were added to his sentence for having stolen—"the State's property"—the prison's dress.

join in large national and international federations, they certainly aid the growth of personal friendly intercourse between all sorts of men scattered in different parts of the globe.

The Alpine Clubs, the *Jagdschutzverein* in Germany, which has over 100,000 members—hunters, educated foresters, zoologists, and simple lovers of Nature—and the International Ornithological Society, which includes zoologists, breeders, and simple peasants in Germany, have the same character. Not only have they done in a few years a large amount of very useful work, which large associations alone could do properly (maps, refuge huts, mountain roads; studies of animal life, of noxious insects, of migrations of birds, and so on), but they create new bonds between men. Two Alpinists of different nationalities who meet in a refuge hut in the Caucasus, or the professor and the peasant ornithologist who stay in the same house, are no more strangers to each other; while the Uncle Toby's Society at Newcastle, which has already induced over 260,000 boys and girls never to destroy birds' nests and to be kind to all animals, has certainly done more for the development of human feelings and of taste in natural science than lots of moralists and most of our schools.

We cannot omit, even in this rapid review, the thousands of scientific, literary, artistic, and educational societies. Up till now, the scientific bodies, closely controlled and often subsidized by the State, have generally moved in a very narrow circle, and they often came to be looked upon as mere openings for getting State appointments, while the very narrowness of their circles undoubtedly bred petty jealousies. Still it is a fact that the distinctions of birth, political parties and creeds are smoothed to some extent by such associations; while in the smaller and remote towns the scientific, geographical, or musical societies, especially those of them which appeal to a larger circle of amateurs, become small centres of intellectual life, a sort of link between the little spot and the wide world, and a place where men of very different conditions meet on a footing of equality. To fully appreciate the value of such centres, one ought to know them, say, in Siberia. As to the countless educational societies which only now begin to break down the State's and the Church's monopoly in education, they are sure to become before long the leading power in that branch. To the "Froebel Unions" we already owe the Kindergarten system; and to a number of formal and informal educational associations we owe the high standard of women's education in Russia, although all the time these societies and groups had to act in strong opposition to a powerful government.[1] As to the various pedagogical societies in Germany, it is well known that they have done the best part in the working out of the modern methods of teaching science in popular schools. In such associatious the teacher finds also his best support. How miserable the overworked and underpaid village teacher would have been without their aid![2]

[1] The Medical Academy for Women (which has given to Russia a large portion of her 700 graduated lady doctors), the four Ladies' Universities (about 1,000 pupils in 1887; closed that year, and reopened in 1895), and the High Commercial School for Women are *entirely* the work of such private societies. To the same societies we owe the high standard which the girls' gymnasia attained since they were opened in the sixties. The 100 gymnasia now scattered over the Empire (over 70,000 pupils), correspond to the High Schools for Girls in this country; all teachers are, however, graduates of the universities.

[2] The *Verein für Verbreitung gemeinnützlicher Kenntnisse*, although

All these associations, societies, brotherhoods, alliances, institutes, and so on, which must now be counted by the ten thousand in Europe alone, and each of which represents an immense amount of voluntary, unambitious, and unpaid or underpaid work—what are they but so many manifestations, under an infinite variety of aspects, of the same ever-living tendency of man towards mutual aid and support? For nearly three centuries men were prevented from joining hands even for literary, artistic, and educational purposes. Societies could only be formed under the protection of the State, or the Church, or as secret brotherhoods, like free-masonry. But now that the resistance has been broken, they swarm in all directions, they extend over all multifarious branches of human activity, they become international, and they undoubtedly contribute, to an extent which cannot yet be fully appreciated, to break down the screens erected by States between different nationalities. Notwithstanding the jealousies which are bred by commercial competition, and the provocations to hatred which are sounded by the ghosts of a decaying past, there is a conscience of international solidarity which is growing both among the leading spirits of the world and the masses of the workers, since they also have conquered the right of international intercourse; and in the preventing of a European war during the last quarter of a century, this spirit has undoubtedly had its share.

The religious charitable associations, which again represent a whole world, certainly must be mentioned in this place. There is not the slightest doubt that

---

it has only 5,500 members, has already opened more than 1,000 public and school libraries, organized thousands of lectures, and published most valuable books.

the great bulk of their members are moved by the same mutual-aid feelings which are common to all mankind. Unhappily the religious teachers of men prefer to ascribe to such feelings a supernatural origin. Many of them pretend that man does not consciously obey the mutual-aid inspiration so long as he has not been enlightened by the teachings of the special religion which they represent, and, with St. Augustin, most of them do not recognize such feelings in the "pagan savage." Moreover, while early Christianity, like all other religions, was an appeal to the broadly human feelings of mutual aid and sympathy, the Christian Church has aided the State in wrecking all standing institutions of mutual aid and support which were anterior to it, or developed outside of it; and, instead of the *mutual aid* which every savage considers as due to his kinsman, it has preached *charity* which bears a character of inspiration from above, and, accordingly, implies a certain superiority of the giver upon the receiver. With this limitation, and without any intention to give offence to those who consider themselves as a body elect when they accomplish acts simply humane, we certainly may consider the immense numbers of religious charitable associations as an outcome of the same mutual-aid tendency.

All these facts show that a reckless prosecution of personal interests, with no regard to other people's needs, is not the only characteristic of modern life. By the side of this current which so proudly claims leadership in human affairs, we perceive a hard struggle sustained by both the rural and industrial populations in order to reintroduce standing institutions of mutual aid and support; and we discover, in all classes of society, a widely-spread movement towards the establishment of an infinite variety of more or less permanent institutions for the same purpose. But when we pass from public life to the private life of the modern individual, we discover another extremely wide world of mutual aid and support, which only passes unnoticed by most sociologists because it is limited to the narrow circle of the family and personal friendship.[1]

Under the present social system, all bonds of union among the inhabitants of the same street or neighbourhood have been dissolved. In the richer parts of the large towns, people live without knowing who are their next-door neighbours. But in the crowded lanes people know each other perfectly, and are continually brought into mutual contact. Of course, petty quarrels go their course, in the lanes as elsewhere; but groupings in accordance with personal affinities grow up, and within their circle mutual aid is practised to an extent of which the richer classes have no idea. If we take, for instance, the children of a poor neighbourhood who play in a street or a churchyard, or on a green, we notice at once

[1] Very few writers in sociology have paid attention to it. Dr. Ihering is one of them, and his case is very instructive. When the great German writer on law began his philosophical work, *Der Zweck im Rechte* ("Purpose in Law"), he intended to analyze "the active forces which call forth the advance of society and maintain it," and to thus give "the theory of the sociable man." He analyzed, first, the egotistic forces at work, including the present wage-system and coercion in its variety of political and social laws; and in a carefully-worked-out scheme of his work he intended to give the last paragraph to the ethical forces—the sense of duty and mutual love—which contribute to the same aim. When he came, however, to discuss the social functions of these two factors, he had to write a second volume, twice as big as the first; and yet he treated only of the *personal* factors which will take in the following pages only a few lines. L. Dargun took up the same idea in *Egoismus und Altruismus in der Nationalökonomie*, Leipzig, 1885, adding some new facts. Büchner's *Love*, and the several paraphrases of it published here and in Germany, deal with the same subject.

that a close union exists among them, notwithstanding the temporary fights, and that that union protects them from all sorts of misfortunes. As soon as a mite bends inquisitively over the opening of a drain—"Don't stop there," another mite shouts out, "fever sits in the hole!" "Don't climb over that wall, the train will kill you if you tumble down! Don't come near to the ditch! Don't eat those berries—poison! you will die!" Such are the first teachings imparted to the urchin when he joins his mates out-doors. How many of the children whose play-grounds are the pavements around "model workers' dwellings," or the quays and bridges of the canals, would be crushed to death by the carts or drowned in the muddy waters, were it not for that sort of mutual support! And when a fair Jack has made a slip into the unprotected ditch at the back of the milk-man's yard, or a cherry-cheeked Lizzie has, after all, tumbled down into the canal, the young brood raises such cries that all the neighbourhood is on the alert and rushes to the rescue.

Then comes in the alliance of the mothers. "You could not imagine" (a lady-doctor who lives in a poor neighbourhood told me lately) "how much they help each other. If a woman has prepared nothing, or could prepare nothing, for the baby which she expected—and how often that happens!—all the neighbours bring something for the new-comer. One of the neighbours always takes care of the children, and some other always drops in to take care of the household, so long as the mother is in bed." This habit is general. It is mentioned by all those who have lived among the poor. In a thousand small ways the mothers support each other and bestow their care upon children that are not their own. Some training—good or bad, let them decide it for themselves—is required in a lady of the richer classes to render her able to pass by a shivering and hungry child in the street without noticing it. But the mothers of the poorer classes have not that training. They cannot stand the sight of a hungry child; they *must* feed it, and so they do. "When the school children beg bread, they seldom or rather never meet with a refusal"—a lady-friend, who has worked several years in Whitechapel in connection with a workers' club, writes to me. But I may, perhaps, as well transcribe a few more passages from her letter:—

"Nursing neighbours, in cases of illness, without any shade of remuneration, is quite general among the workers. Also, when a woman has little children, and goes out for work, another mother always takes care of them.

"If, in the working classes, they would not help each other, they could not exist. I know families which continually help each other—with money, with food, with fuel, for bringing up the little children, in cases of illness, in cases of death.

"The 'mine' and 'thine' is much less sharply observed among the poor than among the rich. Shoes, dress, hats, and so on,—what may be wanted on the spot—are continually borrowed from each other, also all sorts of household things.

"Last winter the members of the United Radical Club had brought together some little money, and began after Christmas to distribute free soup and bread to the children going to school. Gradually they had 1,800 children to attend to. The money came from outsiders, but all the work was done by the members of the club. Some of them, who were out of work, came at four in the morning to wash and to peel the vegetables; five women came at nine or ten (after having done their own household work) for cooking, and stayed till six or seven to wash the dishes. And at meal time, between twelve and half-past one, twenty to thirty workers came in to aid in serving the soup, each one staying what he could spare of his meal time. This lasted for two months. No one was paid."

My friend also mentions various individual cases, of which the following are typical:—

"Annie W. was given by her mother to be boarded by an old person in Wilmot Street. When her mother died, the old woman, who herself was very poor, kept the child without being paid a penny for that. When the old lady died too, the child, who was five years old, was of course neglected during her illness, and was ragged; but she was taken at once by Mrs. S., the wife of a shoemaker, who herself has six children. Lately, when the husband was ill, they had not much to eat, all of them.

"The other day, Mrs. M., mother of six children, attended Mrs. M—g throughout her illness, and took to her own rooms the elder child. . . . But do you need such facts? They are quite general. . . . I know also Mrs. D. (Oval, Hackney Road), who has a sewing machine and continually sews for others, without ever accepting any remuneration, although she has herself five children and her husband to look after. . . . And so on."

For every one who has any idea of the life of the labouring classes it is evident that without mutual aid being practised among them on a large scale they never could pull through all their difficulties. It is only by chance that a worker's family can live its lifetime without having to face such circumstances as the crisis described by the ribbon weaver, Joseph Gutteridge, in his autobiography.[1] And if all do not go to the ground in such cases, they owe it to mutual help. In Gutteridge's case it was an old nurse, miserably poor herself, who turned up at the moment when the family was slipping towards a final catastrophe, and brought in some bread, coal, and bedding, which she had obtained on credit. In other cases, it will be some one else, or the neighbours will take steps to save the family. But without some aid from other poor,

[1] *Light and Shadows in the Life of an Artisan.* Coventry, 1893.

how many more would be brought every year to irreparable ruin![1]

Mr. Plimsoll, after he had lived for some time among the poor, on *7s. 6d.* a week, was compelled to recognize that the kindly feelings he took with him when he began this life "changed into hearty respect and admiration" when he saw how the relations between the poor are permeated with mutual aid and support, and learned the simple ways in which that support is given. After a many years' experience, his conclusion was that "when you come to think of it, such as these men were, so were the vast majority of the working classes."[2] As to bringing up orphans, even by the poorest families, it is so widely-spread a habit, that it may be described as a general rule; thus among the miners it was found, after the two explosions at Warren Vale and at Lund Hill, that "nearly one-third of the men killed, as the respective committees

[1] Many rich people cannot understand how the very poor *can* help each other, because they do not realize upon what infinitesimal amounts of food or money often hangs the life of one of the poorest classes. Lord Shaftesbury had understood this terrible truth when he started his Flowers and Watercress Girls' Fund, out of which loans of one pound, and only occasionally two pounds, were granted, to enable the girls to buy a basket and flowers when the winter sets in and they are in dire distress. The loans were given to girls who had "not a sixpence," but never failed to find some other poor to go bail for them. "Of all the movements I have ever been connected with," Lord Shaftesbury wrote, "I look upon this Watercress Girls' movement as the most successful. . . . It was begun in 1872, and we have had out 800 to 1,000 loans, and have not lost 50*l.* during the whole period. . . . What has been lost—and it has been very little, under the circumstances—has been by reason of death or sickness, not by fraud" (*The Life and Work of the Seventh Earl of Shaftesbury*, by Edwin Hodder, vol. iii. p. 322. London, 1885–86). Several more facts in point in Ch. Booth's *Life and Labour in London*, vol. i.; in Miss Beatrice Potter's "Pages from a Work Girl's Diary" (*Nineteenth Century*, September 1888, p. 310); and so on.

[2] Samuel Plimsoll, *Our Seamen*, cheap edition, London, 1870, p. 110.

can testify, were thus supporting relations other than wife and child." "Have you reflected," Mr. Plimsoll added, "what this is? Rich men, even comfortably-to-do men do this, I don't doubt. But consider the difference." Consider what a sum of one shilling, subscribed by each worker to help a comrade's widow, or *6d.* to help a fellow-worker to defray the extra expense of a funeral, means for one who earns 16*s.* a week and has a wife, and in some cases five or six children to support.[1] But such subscriptions are a general practice among the workers all over the world, even in much more ordinary cases than a death in the family, while aid in work is the commonest thing in their lives.

Nor do the same practices of mutual aid and support fail among the richer classes. Of course, when one thinks of the harshness which is often shown by the richer employers towards their employees, one feels inclined to take the most pessimist view of human nature. Many must remember the indignation which was aroused during the great Yorkshire strike of 1894, when old miners who had picked coal from an abandoned pit were prosecuted by the colliery owners. And, even if we leave aside the horrors of the periods

[1] *Our Seamen*, u.s., p. 110. Mr. Plimsoll added: "I don't wish to disparage the rich, but I think it may be reasonably doubted whether these qualities are so fully developed in them; for, notwithstanding that not a few of them are not unacquainted with the claims, reasonable or unreasonable, of poor relatives, these qualities are not in such constant exercise. Riches seem in so many cases to smother the manliness of their possessors, and their sympathies become, not so much narrowed as—so to speak—stratified: they are reserved for the sufferings of their own class, and also the woes of those above them. They seldom tend downwards much, and they are far more likely to admire an act of courage . . . than to admire the constantly exercised fortitude and the tenderness which are the daily characteristics of a British workman's life"—and of the workmen all over the world as well.

of struggle and social war, such as the extermination of thousands of workers' prisoners after the fall of the Paris Commune—who can read, for instance, revelations of the labour inquest which was made here in the forties, or what Lord Shaftesbury wrote about "the frightful waste of human life in the factories, to which the children taken from the workhouses, or simply purchased all over this country to be sold as factory slaves, were consigned"[1]—who can read that without being vividly impressed by the baseness which is possible in man when his greediness is at stake? But it must also be said that all fault for such treatment must not be thrown entirely upon the criminality of human nature. Were not the teachings of men of science, and even of a notable portion of the clergy, up to a quite recent time, teachings of distrust, despite and almost hatred towards the poorer classes? Did not science teach that since serfdom has been abolished, no one need be poor unless for his own vices? And how few in the Church had the courage to blame the children-killers, while the great numbers taught that the sufferings of the poor, and even the slavery of the negroes, were part of the Divine Plan! Was not Nonconformism itself largely a popular protest against the harsh treatment of the poor at the hand of the Established Church?

With such spiritual leaders, the feelings of the richer classes necessarily became, as Mr. Pimsoll remarked, not so much blunted as "stratified." They seldom went downwards towards the poor, from whom the well-to-do-people are separated by their manner of life, and whom they do not know under their best

[1] *Life of the Seventh Earl of Shaftesbury*, by Edwin Hodder, vol. i. pp. 137–138.

aspects, in their every-day life. But among themselves—allowance being made for the effects of the wealth-accumulating passions and the futile expenses imposed by wealth itself—among themselves, in the circle of family and friends, the rich practise the same mutual aid and support as the poor. Dr. Ihering and L. Dargun are perfectly right in saying that if a statistical record could be taken of all the money which passes from hand to hand in the shape of friendly loans and aid, the sum total would be enormous, even in comparison with the commercial transactions of the world's trade. And if we could add to it, as we certainly ought to, what is spent in hospitality, petty mutual services, the management of other people's affairs, gifts and charities, we certainly should be struck by the importance of such transfers in national economy. Even in the world which is ruled by commercial egotism, the current expression, "We have been harshly treated by that firm," shows that there is also the friendly treatment, as opposed to the harsh, *i.e.* the legal treatment; while every commercial man knows how many firms are saved every year from failure by the friendly support of other firms.

As to the charities and the amounts of work for general well-being which are voluntarily done by so many well-to-do persons, as well as by workers, and especially by professional men, every one knows the part which is played by these two categories of benevolence in modern life. If the desire of acquiring notoriety, political power, or social distinction often spoils the true character of that sort of benevolence, there is no doubt possible as to the impulse coming in the majority of cases from the same mutual-aid feelings. Men who have acquired wealth very often do not find in it the expected satisfaction. Others begin to feel that, whatever economists may say about wealth being the reward of capacity, their own reward is exaggerated. The conscience of human solidarity begins to tell; and, although society life is so arranged as to stifle that feeling by thousands of artful means, it often gets the upper hand; and then they try to find an outcome for that deeply human need by giving their fortune, or their forces, to something which, in their opinion, will promote general welfare.

In short, neither the crushing powers of the centralized State nor the teachings of mutual hatred and pitiless struggle which came, adorned with the attributes of science, from obliging philosophers and sociologists, could weed out the feeling of human solidarity, deeply lodged in men's understanding and heart, because it has been nurtured by all our preceding evolution. What was the outcome of evolution since its earliest stages cannot be overpowered by one of the aspects of that same evolution. And the need of mutual aid and support which had lately taken refuge in the narrow circle of the family, or the slum neighbours, in the village, or the secret union of workers, re-asserts itself again, even in our modern society, and claims its rights to be, as it always has been, the chief leader towards further progress. Such are the conclusions which we are necessarily brought to when we carefully ponder over each of the groups of facts briefly enumerated in the last two chapters.

CONCLUSION

If we take now the teachings which can be borrowed from the analysis of modern society, in connection with the body of evidence relative to the importance of mutual aid in the evolution of the animal world and of mankind, we may sum up our inquiry as follows.

In the animal world we have seen that the vast majority of species live in societies, and that they find in association the best arms for the struggle for life: understood, of course, in its wide Darwinian sense—not as a struggle for the sheer means of existence, but as a struggle against all natural conditions unfavourable to the species. The animal species, in which individual struggle has been reduced to its narrowest limits, and the practice of mutual aid has attained the greatest development, are invariably the most numerous, the most prosperous, and the most open to further progress. The mutual protection which is obtained in this case, the possibility of attaining old age and of accumulating experience, the higher intellectual development, and the further growth of sociable habits, secure the maintenance of the species, its extension, and its further progressive evolution. The unsociable species, on the contrary, are doomed to decay.

Going next over to man, we found him living in clans and tribes at the very dawn of the stone age;

we saw a wide series of social institutions developed already in the lower savage stage, in the clan and the tribe; and we found that the earliest tribal customs and habits gave to mankind the embryo of all the institutions which made later on the leading aspects of further progress. Out of the savage tribe grew up the barbarian village community; and a new, still wider, circle of social customs, habits, and institutions, numbers of which are still alive among ourselves, was developed under the principles of common possession of a given territory and common defence of it, under the jurisdiction of the village folkmote, and in the federation of villages belonging, or supposed to belong, to one stem. And when new requirements induced men to make a new start, they made it in the city, which represented a double network of territorial units (village communities), connected with guilds—these latter arising out of the common prosecution of a given art or craft, or for mutual support and defence.

And finally, in the last two chapters facts were produced to show that although the growth of the State on the pattern of Imperial Rome had put a violent end to all mediæval institutions for mutual support, this new aspect of civilization could not last. The State, based upon loose aggregations of individuals and undertaking to be their only bond of union, did not answer its purpose. The mutual-aid tendency finally broke down its iron rules; it reappeared and reasserted itself in an infinity of associations which now tend to embrace all aspects of life and to take possession of all that is required by man for life and for reproducing the waste occasioned by life.

It will probably be remarked that mutual aid, even though it may represent one of the factors of evolution, covers nevertheless one aspect only of human relations; that by the side of this current, powerful though it may be, there is, and always has been, the other current—the self-assertion of the individual, not only in its efforts to attain personal or caste superiority, economical, political, and spiritual, but also in its much more important although less evident function of breaking through the bonds, always prone to become crystallized, which the tribe, the village community, the city, and the State impose upon the individual. In other words, there is the self-assertion of the individual taken as a progressive element.

It is evident that no review of evolution can be complete, unless these two dominant currents are analyzed. However, the self-assertion of the individual or of groups of individuals, their struggles for superiority, and the conflicts which resulted therefrom, have already been analyzed, described, and glorified from time immemorial. In fact, up to the present time, this current alone has received attention from the epical poet, the annalist, the historian, and the sociologist. History, such as it has hitherto been written, is almost entirely a description of the ways and means by which theocracy, military power, autocracy, and, later on, the richer classes' rule have been promoted, established, and maintained. The struggles between these forces make, in fact, the substance of history. We may thus take the knowledge of the individual factor in human history as granted—even though there is full room for a new study of the subject on the lines just alluded to; while, on the other side, the mutual-aid factor has been hitherto totally lost sight of; it was simply denied, or even scoffed at, by the writers of the present and past generation. It was therefore necessary to show, first of all, the immense part which this factor plays in the evolution of both the animal world and human societies. Only after this has been fully recognized will it be possible to proceed to a comparison between the two factors.

To make even a rough estimate of their relative importance by any method more or less statistical, is evidently impossible. One single war—we all know—may be productive of more evil, immediate and subsequent, than hundreds of years of the unchecked action of the mutual-aid principle may be productive of good. But when we see that in the animal world, progressive development and mutual aid go hand in hand, while the inner struggle within the species is concomitant with retrogressive development; when we notice that with man, even success in struggle and war is proportionate to the development of mutual aid in each of the two conflicting nations, cities, parties, or tribes, and that in the process of evolution war itself (so far as it can go this way) has been made subservient to the ends of progress in mutual aid within the nation, the city or the clan—we already obtain a perception of the dominating influence of the mutual-aid factor as an element of progress. But we see also that the practice of mutual aid and its successive developments have created the very conditions of society life in which man was enabled to develop his arts, knowledge, and intelligence; and that the periods when institutions based on the mutual-aid tendency took their greatest development were also the periods of the greatest progress in arts, industry, and science. In fact, the study of the inner life of the mediæval city and of the ancient Greek cities reveals the fact that the combination of mutual aid, as it was practised within the guild and the Greek clan, with a large initiative which was left to the individual and the group by means of the federative principle, gave to mankind the two greatest periods of its history—the ancient Greek city and the mediæval city periods; while the ruin of the above institutions during the State periods of history, which followed, corresponded in both cases to a rapid decay.

As to the sudden industrial progress which has been achieved during our own century, and which is usually ascribed to the triumph of individualism and competition, it certainly has a much deeper origin than that. Once the great discoveries of the fifteenth century were made, especially that of the pressure of the atmosphere, supported by a series of advances in natural philosophy—and they were made under the mediæval city organization,—once these discoveries were made, the invention of the steam-motor, and all the revolution which the conquest of a new power implied, had necessarily to follow. If the mediæval cities had lived to bring their discoveries to that point, the ethical consequences of the revolution effected by steam might have been different; but the same revolution in technics and science would have inevitably taken place. It remains, indeed, an open question whether the general decay of industries which followed the ruin of the free cities, and was especially noticeable in the first part of the eighteenth century, did not considerably retard the appearance of the steam-engine as well as the consequent revolution in arts. When we consider the astounding rapidity of industrial progress from the twelfth to the fifteenth centuries—in weaving, working of metals, architecture and navigation, and ponder over the scientific discoveries which that industrial progress led to at the end of the fifteenth century—

we must ask ourselves whether mankind was not delayed in its taking full advantage of these conquests when a general depression of arts and industries took place in Europe after the decay of mediæval civilization. Surely it was not the disappearance of the artist-artisan, nor the ruin of large cities and the extinction of intercourse between them, which could favour the industrial revolution; and we know indeed that James Watt spent twenty or more years of his life in order to render his invention serviceable, because he could not find in the last century what he would have readily found in mediæval Florence or Brügge, that is, the artisans capable of realizing his devices in metal, and of giving them the artistic finish and precision which the steam-engine requires.

To attribute, therefore, the industrial progress of our century to the war of each against all which it has proclaimed, is to reason like the man who, knowing not the causes of rain, attributes it to the victim he has immolated before his clay idol. For industrial progress, as for each other conquest over nature, mutual aid and close intercourse certainly are, as they have been, much more advantageous than mutual struggle.

However, it is especially in the domain of ethics that the dominating importance of the mutual-aid principle appears in full. That mutual aid is the real foundation of our ethical conceptions seems evident enough. But whatever the opinions as to the first origin of the mutual-aid feeling or instinct may be—whether a biological or a supernatural cause is ascribed to it—we must trace its existence as far back as to the lowest stages of the animal world; and from these stages we can follow its uninterrupted evolution, in opposition to a number of contrary agencies, through all degrees of human development, up to the present times. Even the new religions which were born from time to time—always at epochs when the mutual-aid principle was falling into decay in the theocracies and despotic States of the East, or at the decline of the Roman Empire—even the new religions have only reaffirmed that same principle. They found their first supporters among the humble, in the lowest, downtrodden layers of society, where the mutual-aid principle is the necessary foundation of every-day life; and the new forms of union which were introduced in the earliest Buddhist and Christian communities, in the Moravian brotherhoods and so on, took the character of a return to the best aspects of mutual aid in early tribal life.

Each time, however, that an attempt to return to this old principle was made, its fundamental idea itself was widened. From the clan it was extended to the stem, to the federation of stems, to the nation, and finally—in ideal, at least—to the whole of mankind. It was also refined at the same time. In primitive Buddhism, in primitive Christianity, in the writings of some of the Mussulman teachers, in the early movements of the Reform, and especially in the ethical and philosophical movements of the last century and of our own times, the total abandonment of the idea of revenge, or of "due reward"—of good for good and evil for evil—is affirmed more and more vigorously. The higher conception of "no revenge for wrongs," and of freely giving more than one expects to receive from his neighbours, is proclaimed as being the real principle of morality—a principle superior to mere equivalence, equity, or justice, and more conducive to happiness. And man is appealed to to be guided in his acts, not merely by love, which is always personal, or at the best tribal, but by the perception of his oneness with each human being. In the practice of mutual aid, which we can retrace to the earliest beginnings of evolution, we thus find the positive and undoubted origin of our ethical conceptions; and we can affirm that in the ethical progress of man, mutual support—not mutual struggle—has had the leading part. In its wide extension, even at the present time, we also see the best guarantee of a still loftier evolution of our race.

APPENDIX

I.—SWARMS OF BUTTERFLIES, DRAGON-FLIES, ETC.

(To p. 10.)

M. C. PIEPERS has published in *Natuurkunding Tijdschrift voor Neederlandsch Indië*, 1891, Deel L. p. 198 (analyzed in *Naturwissenschaftliche Rundschau*, 1891, vol. vi. p. 573), interesting researches into the mass-flights of butterflies which occur in Dutch East India, seemingly under the influence of great draughts occasioned by the west monsoon. Such mass-flights usually take place in the first months after the beginning of the monsoon, and it is usually individuals of both sexes of *Catopsilia* (*Callidryas*) *crocale*, Cr., which join in it, but occasionally the swarms consist of individuals belonging to three different species of the genus *Euphœa*. Copulation seems also to be the purpose of such flights. That these flights are not the result of concerted action but rather a consequence of imitation, or of a desire of following all others, is, of course, quite possible.

Bates saw, on the Amazon, the yellow and the orange *Callidryas* "assembling in densely packed masses, sometimes two or three yards in circumference, their wings all held in an upright position, so that the beach looked as though variegated with beds of crocuses." Their migrating columns, crossing the river from north to south, "were uninterrupted, from an early hour in the morning till sunset" (*Naturalist on the Amazon*, p. 131).

Dragon-flies, in their long migrations across the Pampas, come together in countless numbers, and their immense swarms contain individuals belonging to different species (Hudson, *Naturalist on the La Plata*, pp. 130 *seq.*).

The grasshoppers (*Zoniopoda tarsata*) are also eminently gregarious (Hudson, *l. c.* p. 125).

II.—THE ANTS.

(To p. 13.)

Pierre Huber's *Les fourmis indigènes* (Genève, 1810), of which a cheap edition was issued in 1861 by Cherbuliez, in the *Bibliothèque Genevoise*, and of which translations ought to be circulated in cheap editions in every language, is not only the best work on the subject, but also a model of really scientific research. Darwin was quite right in describing Pierre Huber as an even greater naturalist than his father. This book ought to be read by every young naturalist, not only for the facts it contains but as a lesson in the methods of research. The rearing of ants in artificial glass nests, and the test experiments made by subsequent explorers, including Lubbock, will all be found in Huber's admirable little work. Readers of the books of Forel and Lubbock are, of course, aware that both the Swiss professor and the British writer began their work in a critical mood, with the intention of disproving Huber's assertions concerning the admirable mutual-aid instincts of the ants; but that after a careful investigation they could only confirm them. However, it is unfortunately characteristic of human nature gladly to believe any affirmation concerning men being able to change at will the action of the forces of Nature, but to refuse to admit well-proved scientific facts tending to reduce the distance between man and his animal brothers.

Mr. Sutherland (*Origin and Growth of Moral Instinct*) evidently began his book with the intention of proving that all moral feelings have originated from parental care and familial love, which both appeared only in warm-blooded animals; consequently he tries to minimize the importance of sympathy and co-operation among ants. He quotes Büchner's book, *Mind in Animals*, and knows Lubbock's experiments. As to the works of Huber and Forel, he dismisses them in the following sentence; "but they [Büchner's instances of sympathy among ants] are all, or mostly all, marred by a certain air of sentimentalism . . . which renders them better suited for school books than for cautious works of science, and *the same is to be remarked* [italics are mine] of some of Huber's and Forel's best-known *anecdotes*" (vol. i. p. 298).

Mr. Sutherland does not specify which "anecdotes" he means, but it seems to me that he could never have had the opportunity of perusing the works of Huber and Forel. Naturalists who know these works find no "anecdotes" in them.

The recent work of Professor Gottfried Adlerz on the ants in Sweden, *Myrmecologiska Studier: Svenska Myror och des Lefnadsförhållanden* (*Bihang till Svenska Akademiens Handlingar*, Bd. xi. No. 18, 1886), may be mentioned in this place. It hardly need be said that all the observations of Huber and Forel concerning the mutual-aid life of ants, including the one concerning the sharing of food, felt to be so striking by those who previously had paid no attention to the subject, are fully confirmed by the Swedish professor (pp. 136–137).

Professor G. Adlerz gives also very interesting experiments to prove what Huber had already observed; namely, that ants from two different nests do not always attack each other. He has made one of his experiments with the ant, *Tapinoma erraticum*. Another was made with the common *Rufa* ant. Taking a whole nest in a sack, he emptied it at a distance of six feet from another nest. There was no battle, but the ants of the second nest began to carry the pupæ of the former. As a rule, when Professor Adlerz brought together workers with their pupæ, both taken from different nests, there was no battle; but if the workers were without their pupæ, a battle ensued (pp. 185–186).

He also completes Forel's and MacCook's observations about the "nations" of ants, composed of many nests, and, taking his own estimates, which brought him to take an average of 300,000 *Formica exsecta* ants in each nest, he concludes that such "nations" may reach scores and even hundreds of millions of inhabitants.

Maeterlinck's admirably written book on bees, although it contains no new observations, would be very useful, if it were less marred with metaphysical "words."

III.—NESTING ASSOCIATIONS.

(To p. 35.)

Audubon's Journals (*Audubon and his Journals*, New York, 1898), especially those relating to his life on the coasts of Labrador and the St. Lawrence river in the thirties, contain excellent descriptions of the nesting associations of aquatic birds. Speaking of "The Rock," one of the Magdalene or Amherst Islands, he wrote:—"At eleven I could distinguish its top plainly from the deck, and thought it covered with snow to the depth of several feet; this appearance existed on every portion of the flat, projecting shelves." But it was not snow: it was gannets, all calmly seated on their eggs or newly-hatched brood—their heads all turned windwards, almost touching each other, and in regular lines. The air above, for a hundred yards and for some distance round the rock, "was filled with gannets on the wing, as if a heavy fall of snow was directly above us." Kittiwake gulls and foolish guillemots bred on the same rock (*Journals*, vol. i. pp. 360–363).

In sight of Anticosti Island, the sea "was literally covered with foolish guillemots and with razor-billed auks (*Alca torva*)." Further on, the air was filled with velvet ducks. On the rocks of the Gulf, the herring gulls, the terns (great, Arctic, and probably Foster's), the *Tringa pusilla*, the sea-gulls, the auks, the Scoter ducks, the wild geese (*Anser canadensis*), the red-breasted merganser, the cormorants, etc., were all breeding. The sea-gulls were extremely abundant there; "they are for ever harassing every other bird, sucking their eggs and devouring their young;" "they take here the place of eagles and hawks."

On the Missouri, above Saint Louis, Audubon saw, in 1843, vultures and eagles nesting in colonies. Thus he mentioned "long lines of elevated shore, surmounted by stupendous rocks of limestone, with many curious holes in them, where we saw vultures and eagles enter towards dusk"—that is, Turkey buzzards (*Cathartes aura*) and bald eagles (*Haliaëtus leucocephalus*), E. Couës remarks in a footnote (vol. i. p. 458).

One of the best breeding-grounds along the British shores are the Farne Islands, and one will find in Charles Dixon's work, *Among the Birds in Northern Shires*, a lively description of these grounds, where scores of thousands of gulls, terns, eider-ducks, cormorants, ringed plovers, oyster-catchers, guillemots, and puffins come together every year. "On approaching some of the islands the first impression is that this gull (the lesser black-backed gull) monopolizes the whole of the ground, as it occurs in such vast abundance. The air seems full of them, the ground and bare rocks are crowded; and as our boat finally grates against the rough beach and we eagerly jump ashore all becomes noisy excitement—a perfect babel of protesting cries that is persistently kept up until we leave the place" (p. 219).

IV.—SOCIABILITY OF ANIMALS.

(To p. 42.)

That the sociability of animals was greater when they were less hunted by man, is confirmed by many facts showing that those animals who now live isolated in countries inhabited by man continue to live in herds in uninhabited regions. Thus on the waterless plateau deserts of Northern Thibet Prjevalsky found bears living in societies. He mentions numerous "herds of yaks, khulans, antelopes, and even bears." The latter, he says, feed upon the extremely numerous small rodents, and are so numerous that, "as the natives assured me, they have found a hundred or a hundred and fifty of them asleep in the same cave" (*Yearly Report* of the Russian Geographical Society for 1885, p. 11; Russian). Hares (*Lepus Lehmani*) live in large societies in the Transcaspian territory (N. Zarudnyi, *Recherches zoologiques dans la contrée Transcaspienne*, in *Bull. Soc. Natur. Moscou*, 1889, 4). The small Californian foxes, who, according to E. S. Holden, live round the Lick observatory "on a mixed diet of Manzanita berries and astronomers' chickens"

(*Nature*, Nov. 5, 1891), seem also to be very sociable.

Some very interesting instances of the love of society among animals have lately been given by Mr. C. J. Cornish (*Animals at Work and Play*, London, 1896). All animals, he truly remarks, hate solitude. He gives also an amusing instance of the habit of the prairie dogs of keeping sentries. It is so great that they always keep a sentinel on duty, even at the London Zoological Garden, and in the Paris Jardin d'Acclimatation (p. 46).

Professor Kessler was quite right in pointing out that the young broods of birds, keeping together in autumn, contribute to the development of feelings of sociability. Mr. Cornish (*Animals at Work and Play*) has given several examples of the plays of young mammals, such as, for instance, lambs playing at "follow my leader," or at "I'm the king of the castle," and their love of steeplechases; also the fawns playing a kind of "cross-touch," the touch being given by the nose. We have, moreover, the excellent work by Karl Gross, *The Play of Animals.*

V.—CHECKS TO OVER-MULTIPLICATION.

(To p. 72.)

Hudson, in his *Naturalist on the La Plata* (Chapter III.), has a very interesting account of a sudden increase of a species of mice and of the consequences of that sudden "wave of life."

"In the summer of 1872–73," he writes, "we had plenty of sunshine, with frequent showers, so that the hot months brought no dearth of wild flowers, as in most years." The season was very favourable for mice, and "these prolific little creatures were soon so abundant that the dogs and the cats subsisted almost exclusively on them. Foxes, weasels and opossums fared sumptuously; even the insectivorous armadillo took to mice-hunting." The fowls became quite rapacious, "while the sulphur tyrant-birds (*Pitangus*) and the *Guira* cuckoos preyed on nothing but mice." In the autumn, countless numbers of storks and of short-eared owls made their appearance, coming also to assist at the general feast. Next came a winter of continued drought; the dry grass was eaten, or turned to dust; and the mice, deprived of cover and food, began to die out. The cats sneaked back to the houses; the short-eared owls—a wandering species—left; while the little burrowing owls became so reduced as scarcely to be able to fly, "and hung about the houses all day long on the look-out for some stray morsel of food." Incredible numbers of sheep and cattle perished the same winter, during a month of cold that followed the drought. As to the mice, Hudson makes the remark that "scarcely a hard-pressed remnant remains after the great reaction, to continue the species."

This illustration has an additional interest in its showing how, on flat plains and plateaus, the sudden increase of a species immediately attracts enemies from other parts of the plains, and how species unprotected by their social organization must necessarily succumb before them.

Another excellent illustration in point is given by the same author from the Argentine Republic. The coypù (*Myopotamus coypù*) is there a very common rodent—a rat in shape, but as large as an otter. It is aquatic in its habits and very sociable. "Of an evening," Hudson writes, "they are all out swimming and playing in the water, conversing together in strange tunes, which sound like the moans and cries of wounded and suffering men. The coypù, which has a fine fur under the long coarse hair, was largely exported to Europe; but some sixty years ago the Dictator Rosas issued a decree prohibiting the hunting of this animal. The result was that the animals increased and multiplied exceedingly, and, abandoning their aquatic habits, they became terrestrial and migratory, and swarmed everywhere in search of food. Suddenly a mysterious malady fell on them, from which they quickly perished, and became almost extinct" (p. 12).

Extermination by man on the one side, and contagious diseases on the other side, are thus the main checks which keep the species down—not competition for the means of existence, which may not exist at all.

Facts, proving that regions enjoying a far more congenial climate than Siberia are equally under-populated, could be produced in numbers. But in Bates' well-known work we find the same remark concerning even the shores of the Amazon river.

"There is, in fact," Bates wrote, "a great variety of mammals, birds and reptiles, but they are widely scattered and all excessively shy of man. The region is so extensive and uniform in the forest-clothing of its surface, that it is only at long intervals that animals are seen in abundance, where some particular spot is found which is more attractive than the others" (*Naturalist on the Amazon*, 6th ed., p. 31).

This fact is the more striking as the Brazilian fauna, which is poor in mammals, is not poor at all in birds, and the Brazilian forests afford ample food for birds, as may be seen from a quotation, already given on a previous page, about birds' societies. And yet, the forests of Brazil, like those of Asia and Africa, are not over-populated, but rather under-populated. The same is true concerning the pampas of South America, about which W. H. Hudson remarks that it is really astonishing that only one small ruminant should be found on this immense grassy area, so admirably suited to herbivorous quadrupeds. Millions of sheep, cattle and horses, introduced by man, graze now, as is known, upon a portion of these prairies. Land-birds on the pampas are also few in species and in numbers.

VI.—ADAPTATIONS TO AVOID COMPETITION.

(To p. 75.)

Numerous examples of such adaptations can be found in the works of all field-naturalists. One of them, very interesting, may be given in the hairy armadillo, of which W. H. Hudson says, that "it has struck a line for itself, and consequently thrives, while its congeners are fast disappearing. Its food is most varied. It preys on all kinds of insects, discovering worms and larvæ several inches beneath the surface. It is fond of eggs and fledglings; it feeds on carrion as readily as a vulture; and, failing animal food, it subsists on vegetable diet—clover, and even grains of maize. Therefore, when other animals are starving, the hairy armadillo is always fat and vigorous" (*Naturalist on the La Plata*, p. 71).

The adaptivity of the lapwing makes it a species of which the range of extension is very wide. In England, it "makes itself at home on arable land as readily as in wilder areas." Ch. Dixon says in his *Birds of Northern Shires* (p. 67), "Variety of food is still more the rule with the birds of prey." Thus, for instance, we learn from the same author (pp. 60, 65), "that the hen harrier of the British moors feeds not only on small birds, but also on moles and mice, and on frogs, lizards and insects, while most of the smaller falcons subsist largely on insects."

The very suggestive chapter which W. H. Hudson gives to the family of the South American tree-creepers, or woodhewers, is another excellent illustration of the ways in which large portions of the animal population avoid competition, while at the same time they succeed in becoming very numerous in a given region, without being possessed of any of the weapons usually considered as essential in the struggle for existence. The above family covers an immense range, from South Mexico to Patagonia, and no fewer than 290 species, referable to about 46 genera, are already known from this family, the most striking feature of which is the great diversity of habits of its members. Not only the different genera and the different species possess habits peculiarly their own, but even the same species is often found to differ in its manner of life in different localities. "Some species of *Xenops* and *Magarornis*, like woodpeckers, climb vertically on tree-trunks in search of insect prey, but also, like tits, explore the smaller twigs and foliage at the extremity of the branches; so that the whole tree, from the root to its topmost foliage, is hunted over by them. The *Sclerurus*, although an inhabitant of the darkest forest, and provided with sharply-curved claws, never seeks its food on trees, but exclusively on the ground, among the decaying fallen leaves; but, strangely enough, when alarmed, it flies to the trunk of the nearest tree, to which it clings in a vertical position, and, remaining silent and motionless, escapes observation by means of its dark protective colour." And so on. In their nesting habits they also vary immensely. Thus, in one single genus, three species build an oven-shaped clay-nest, a fourth builds a nest of sticks in the trees, and a fifth burrows in the side of a bank, like a kingfisher.

Now, this extremely large family, of which Hudson says that "every portion of the South American continent is occupied by them; for there is really no climate, and no kind of soil or vegetation, which does not possess its appropriate species, belongs"—to use his own words—"to the most defenceless of birds." Like the ducks which were mentioned by Syevertsoff (see in the text), they display no powerful beak or claws; "they are timid, unresisting creatures, without strength or weapons; their movements are less quick and vigorous than those of other kinds, and their flight is exceedingly feeble." But they possess—both Hudson and Asara observe—"the social disposition in an eminent degree," although "the social habit is kept down in them by the conditions of a life which makes solitude necessary." They cannot make those large breeding associations which we see in the sea-birds, because they live on the tree-insects, and they must carefully explore separately every tree—which they do in a most business-like way; but they continually call each other in the woods, "conversing with one another over long distances;" and they associate in those "wandering bands" which are well known from Bates' picturesque description, while Hudson was led to believe "that everywhere in South America the Dendrocolaptidæ are the first in combining to act in concert, and that the birds of other families follow their march and associate with them, knowing from experience that a rich harvest may be reaped." It hardly need be added that Hudson pays them also a high compliment concerning their intelligence. Sociability and intelligence always go hand in hand.

VII.—THE ORIGIN OF THE FAMILY.

(To p. 86.)

At the time when I wrote the chapter inserted in the text, a certain accord seemed to have been established amongst anthropologists concerning the relatively late appearance, in the institutions of men, of the patriarchal family, such as we know it among the Hebrews, or in Imperial Rome. However, works have been published since, in which the ideas promulgated by Bachofen and MacLennan, systematized especially by Morgan, and further developed and confirmed by Post, Maxim Kovalevsky, and Lubbock, were contested—the most important of such works being by the Danish Professor, C. N. Starcke (*Primitive Family*, 1889), and by the Helsingfors Professor, Edward Westermarck (*The History of Human Marriage*, 1891; 2nd ed. 1894). The same has happened with this question of primitive marriage institutions as it happened with the question of the primitive land-ownership institutions. When the ideas of Maurer and Nasse on the village community, developed by quite a school of gifted explorers, and those of all modern anthropologists upon the primitively communistic constitution of the clan had nearly won general acceptance—they called forth the appearance of such works as those of Fustel de Coulanges in France, Frederic Seebohm in England, and several others, in which an attempt was made—with more brilliancy than real depth of investigation—to undermine these ideas and to cast a doubt upon the conclusions arrived at by modern research (see Prof. Vinogradov's Preface to his remarkable work, *Villainage in England*). Similarly, when the ideas about the non-existence of the family at the early tribal stage of mankind began to be accepted by most anthropologists and students of ancient law, they necessarily called forth such works as those of Starcke and Westermarck, in which man was represented, in accordance with the Hebrew tradition, as having started with the family, evidently patriarchal, and never having passed through the stages described by MacLennan, Bachofen, or Morgan. These works, of which the brilliantly-written *History of Human Marriage* has especially been widely read, have undoubtedly produced a certain effect: those who have not had the opportunity of reading the bulky volumes related to the controversy became hesitating; while some anthropologists, well acquainted with the matter, like the French Professor Durkheim, took a conciliatory, but somewhat undefined attitude.

For the special purpose of a work on Mutual Aid, this controversy may be irrelevant. The fact that men have lived in *tribes* from the earliest stages of mankind, is not contested, even by those who feel shocked at the idea that man may have passed through a stage when the family as we understand it did not exist. The subject, however, has its own interest and deserves to be mentioned, although it must be remarked that a volume would be required to do it full justice.

When we labour to lift the veil that conceals from us ancient institutions, and especially such institutions as have prevailed at the first appearance of beings of the human type, we are bound—in the necessary absence of direct testimony—to accomplish a most painstaking work of tracing backwards every institution, carefully noting even its faintest traces in habits, customs, traditions, songs, folk-lore, and so on; and then, combining the separate results of each of these separate studies, to mentally reconstitute the society which would answer to the co-existence of all these institutions. One can consequently understand what a formidable array of facts, and what a vast number of minute studies of particular points is required to come to any safe conclusion. This is exactly what one finds in the monumental work of Bachofen and his followers, but fails to find in the works of the other school. The mass of facts ransacked by Prof. Westermarck is undoubtedly great enough, and his work is certainly very valuable as a criticism; but it hardly will induce those who know the works of Bachofen, Morgan, MacLennan, Post, Kovalevsky, etc., in the originals, and are acquainted with the village-community school, to change their opinions and accept the patriarchal family theory.

Thus the arguments borrowed by Westermarck from the familiar habits of the primates have not, I dare say, the value which he attributes to them. Our knowledge about the family relations amongst the sociable species of monkeys of our own days is extremely uncertain, while the two unsociable species of orang-outan and gorilla must be ruled out of discussion, both being evidently, as I have indicated in the text, decaying species. Still less do we know about the relations which existed between males and females amongst the primates towards the end of the Tertiary period. The species which lived then are probably all extinct, and we have not the slightest idea as to which of them was the ancestral form which Man sprung from. All we can say with any approach to probability is, that various family and tribe relations must have existed in the different ape species, which were extremely numerous at that time; and that great changes must have taken place since in the habits of the primates, similarly to the changes that took place, even within the last two centuries, in the habits of many other mammal species.

The discussion must consequently be limited entirely to human institutions; and in the minute discussion of each separate trace of each early institution, *in connection with all that we know about every other institution of the same people or the same tribe,* lies the main force of the argument of the school which maintains that the patriarchal family is an institution of a relatively late origin.

There is, in fact, *quite a cycle of institutions* amongst primitive men, which become fully comprehensible if we accept the ideas of Bachofen and Morgan, but are utterly incomprehensible otherwise. Such are: the communistic life of the clan, so long as it was not split up into separate paternal families; the life in *long houses,* and in *classes* occupying separate long houses according to the age and stage of initiation of the youth (M. Maclay, H. Schurz); the restrictions to personal accumulation of property of which several illustrations are given above, in the text; the fact that women taken from another tribe belonged to the whole tribe before becoming private property; and many similar institutions analyzed by Lubbock. This wide cycle of institutions, which fell into decay and finally disappeared in the village-community phase of human development, stand in perfect accord with the "tribal marriage" theory; but they are mostly left unnoticed by the followers of the patriarchal family school. This is certainly not the proper way of discussing the problem. Primitive men have not several superposed or juxtaposed institutions as we have now. They have but *one* institution, the clan, which embodies *all* the mutual relations of the members of the clan. Marriage-relations and possession-relations are clan-relations. And the last that we might expect from the defenders of the patriarchal family theory would be to show us how the just mentioned cycle of institutions (which disappear later on) could have existed in an agglomeration of men living under a system contradictory of such institutions —the system of separate families governed by the *pater familias.*

Again, one cannot recognize scientific value in the

way in which certain serious difficulties are set aside by the promoters of the patriarchal family theory. Thus, Morgan has proved by a considerable amount of evidence that a strictly-kept "classificatory group system" exists with many primitive tribes, and that all the individuals of the same category address each other as if they were brothers and sisters, while the individuals of a younger category will address their mothers' sisters as mothers, and so on. To say that this must be a simple *façon de parler*—a way of expressing respect to age—is certainly an easy method of getting rid of the difficulty of explaining, why this special mode of expressing respect, and not some other, has prevailed among so many peoples of different origin, so as to survive with many of them up to the present day? One may surely admit that *ma* and *pa* are the syllables which are easiest to pronounce for a baby, but the question is—Why this part of "baby language" is used by full-grown people, and is applied to a certain strictly-defined category of persons? Why, with so many tribes in which the mother and her sisters are called *ma*, the father is designated by *tiatia* (similar to *diadia*—uncle), *dad*, *da* or *pa?* Why the appellation of mother given to maternal aunts is supplanted later on by a separate name? And so on. But when we learn that with many savages the mother's sister takes as responsible a part in bringing up a child as the mother itself, and that, if death takes away a beloved child, the other "mother" (the mother's sister) will sacrifice herself to accompany the child in its journey into the other world—we surely see in these names something much more profound than a mere *façon de parler*, or a way of testifying respect. The more so when we learn of the existence of quite a cycle of survivals (Lubbock, Kovalevsky, Post have fully discussed them), all pointing in the same direction. Of course it may be said that kinship is reckoned on the maternal side "because the child remains more with its mother," or we may explain the fact that a man's children by several wives of different tribes belong to their mothers' clans in consequence of the savages' "ignorance of physiology;" but these are not arguments even approximately adequate to the seriousness of the questions involved—especially when it is known that the obligation of bearing the mother's name implies belonging to the mother's clan in all respects: that is, involves a right to all the belongings of the maternal clan, as well as the right of being protected by it, never to be assailed by any one of it, and the duty of revenging offences on its behalf.

Even if we were to admit for a moment the satisfactory nature of such explanations, we should soon find out that a separate explanation has to be given for each category of such facts—and they are very numerous. To mention but a few of them, there is: the division of clans into classes, at a time when there is no division as regards property or social condition; exogamy and all the consequent customs enumerated by Lubbock; the blood covenant and a series of similar customs intended to testify the unity of descent; the appearance of family gods subsequent to the existence of clan gods; the exchange of wives which exists not only with Eskimos in times of calamity, but is also widely spread among many other tribes of a quite different origin; the looseness of nuptial ties the lower we descend in civilization; the compound marriages—several men marrying one wife who belongs to them in turns; the abolition of the marriage restrictions during festivals, or on each fifth, sixth, etc., day; the cohabitation of families in "long houses;" the obligation of rearing the orphan falling, even at a late period, upon the maternal uncle; the considerable number of transitory forms showing the gradual passage from maternal descent to paternal descent; the limitation of the number of children by the clan—not by the family—and the abolition of this harsh clause in times of plenty; family restrictions coming after the clan restrictions; the sacrifice of the old relatives to the tribe; the tribal *lex talionis* and many other habits and customs which become a "family matter" only when we find the family, in the modern sense of the word, finally constituted; the nuptial and pre-nuptial ceremonies of which striking illustrations may be found in the work of Sir John Lubbock, and of several modern Russian explorers; the absence of marriage solemnities where the line of descent is matriarchal, and the appearance of such solemnities with tribes following the paternal line of descent—all these and many others[1] showing that, as Durkheim remarks, marriage proper "is only tolerated and prevented by antagonist forces;" the destruction at the death of the individual of what belonged to him personally; and finally, all the formidable array of survivals,[2] myths (Bachofen and his many followers), folk-lore, etc., all telling in the same direction.

Of course, all this does not prove that there was a period when woman was regarded as superior to man, or was the "head" of the clan; this is a quite distinct matter, and my personal opinion is that no such period has ever existed; nor does it prove that there was a time when no tribal restrictions to the union of sexes existed—this would have been absolutely contrary to all known evidence. But when all the facts lately brought to light are considered in their mutual dependency, it is impossible not to recognize that if isolated couples, with their children, have possibly existed even in the primitive clan, these incipient

[1] See *Marriage Customs in many Lands*, by H. N. Hutchinson, London, 1897.

[2] Many new and interesting forms of these have been collected by Wilhelm Rudeck, *Geschichte der öffentlichen Sittlichkeit in Deutschland*, analyzed by Durckheim in *Annuaire Sociologique*, ii. 312.

families were *tolerated exceptions only*, not the institution of the time.

VIII.—DESTRUCTION OF PRIVATE PROPERTY ON THE GRAVE.

(To p. 99.)

In a remarkable work, *The Religious Systems of China*, published in 1892-97 by J. M. de Groot at Leyden, we find the confirmation of this idea. There was in China (as elsewhere) a time when all personal belongings of a dead person were destroyed on his tomb—his mobiliary goods, his chattels, his slaves, and even friends and vassals, and of course his widow. It required a strong reaction against this custom on behalf of the moralists to put an end to it. With the gipsies in England the custom of destroying all chattels on the grave has survived up to the present day. All the personal property of the gipsy queen who died a few years ago was destroyed on her grave. Several newspapers mentioned it at that time.

IX.—THE "UNDIVIDED FAMILY."

(To p. 124.)

A number of valuable works on the South Slavonian *Zadruga*, or "compound family," compared to other forms of family organization, have been published since the above was written; namely, by Ernest Miler (*Jahrbuch der Internationaler Vereinung für vergleichende Rechtswissenschaft und Volkswirthschaftslehre*, 1897), and I. E. Geszow's *Zadruga in Bulgaria*, and

Zadruga-Ownership and Work in Bulgaria (both in Bulgarian). I must also mention the well-known study of Bogisic (*De la forme dite 'inokosna' de la famille rurale chez les Serbes et les Croates*, Paris, 1884), which has been omitted in the text.

X.—THE ORIGIN OF THE GUILDS.

(To p. 176.)

The origin of the guilds has been the subject of many controversies. There is not the slightest doubt that craft-guilds, or "colleges" of artisans, existed in ancient Rome. It appears, indeed, from a passage in Plutarch that Numa legislated about them. "He divided the people," we are told, "into trades . . . ordering them to have brotherhoods, festivals, and meetings, and indicating the worship they had to accomplish before the gods, according to the dignity of each trade." It is almost certain, however, that it was not the Roman king who invented, or instituted, the trade-colleges—they had already existed in ancient Greece; in all probability, he simply submitted them to royal legislation, just as Philippe le Bel, fifteen centuries later, submitted the trades of France, much to their detriment, to royal supervision and legislation. One of the successors of Numa, Servius Tullius, also is said to have issued some legislation concerning the colleges.[1]

Consequently, it was quite natural that historians should ask themselves whether the guilds which took

[1] A Servio Tullio populus romanus relatus in censum, digestus in classes, curiis atque collegiis distributus (E. Martin-Saint-Léon, *Histoire des corporations de métiers depuis leurs origines jusqu'à leur suppression en* 1791, etc., Paris, 1897).

such a development in the twelfth, and even the tenth and the eleventh centuries, were not revivals of the old Roman "colleges"—the more so as the latter, as seen from the above quotation, quite corresponded to the mediæval guild.[1] It is known, indeed, that corporations of the Roman type existed in Southern Gaul down to the fifth century. Besides, an inscription found during some excavations in Paris shows that a corporation of Lutetia *nautæ* existed under Tiberius; and in the chart given to the Paris "water-merchants" in 1170, their rights are spoken of as existing *ab antiquo* (same author, p. 51). There would have been, therefore, nothing extraordinary, had corporations been maintained in early mediæval France after the barbarian invasions.

However, even if as much must be granted, there is no reason to maintain that the Dutch corporations, the Norman guilds, the Russian *artéls*, the Georgian *amkari*, and so on, necessarily have had also a Roman, or even a Byzantine origin. Of course, the intercourse between the Normans and the capital of the East-Roman Empire was very active, and the Slavonians (as has been proved by Russian historians, and especially by Rambaud) took a lively part in that intercourse. So, the Normans and the Russians may have imported the Roman organization of trade-corporations into their respective lands. But when we see that the *artél* was the very essence of the every-day life of all the Russians, as early as the tenth century, and that this *artél*, although no sort of legislation has ever regulated its life till modern times, has the very same features as the Roman college and the Western guild, we are still more inclined to consider the eastern guild as having an even more ancient origin than the Roman college. Romans knew well, indeed, that

[1] The Roman *sodalitia*, so far as we may judge (same author, p. 9), corresponded to the Kabyle *çofs*.

their *sodalitia* and *collegia* were "what the Greeks called *hetairiai*" (Martin-Saint-Léon, p. 2), and from what we know of the history of the East, we may conclude, with little probability of being mistaken, that the great nations of the East, as well as Egypt, also have had the same guild organization. The essential features of this organization remain the same wherever we may find them. It is a union of men carrying on the same profession or trade. This union, like the primitive clan, has its own gods and its own worship, always containing some mysteries, specific to each separate union; it considers all its members as *brothers and sisters*—possibly (at its beginnings) with all the consequences which such a relationship implied in the gens, or, at least, with ceremonies that indicated or symbolized the clan relations between brother and sister; and finally, all the obligations of mutual support which existed in the clan, exist in this union; namely, the exclusion of the very possibility of a murder within the brotherhood, the clan responsibility before justice, and the obligation, in case of a minor dispute, of bringing the matter before the judges, or rather the arbiters, of the guild brotherhood. The guild—one may say—is thus modelled upon the clan.

Consequently, the same remarks which are made in the text concerning the origin of the village community, apply, I am inclined to think, equally to the guild, the *artél*, and the craft- or neighbour-brotherhood. When the bonds which formerly connected men in their clans were loosened in consequence of migrations, the appearance of the paternal family, and a growing diversity of occupations—a new *territorial bond* was worked out by mankind in the shape of the village community; and another bond—an *occupation bond*—was worked out in an imaginary brotherhood—*the imaginary clan*, which was represented: between two men, or a few men, by the "mixture-of-blood brotherhood" (the Slavonian *pobratimstvo*), and between a greater number of men of different origin, *i.e.* originated from different clans, inhabiting the same village or town (or even different villages or towns)—the *phratry*, the *hetairiai*, the *amkari*, the *artél*, the guild.[1]

As to the idea and the form of such an organization, its elements were already indicated from the savage period downwards. We know indeed that in the clans of all savages there are separate secret organizations of warriors, of witches, of young men, etc.—craft mysteries, in which knowledge concerning hunting or warfare is transmitted; in a word, "clubs," as Miklukho-Maclay described them. These "mysteries" were, in all probability, the prototypes of the future guilds.[2]

[1] It is striking to see how distinctly this very idea is expressed in the well-known passage of Plutarch concerning Numa's legislation of the trade-colleges:—"And through this," Plutarch wrote, "he was the first to banish from the city this spirit which led people to say: 'I am a Sabine,' or 'I am a Roman,' or 'I am a subject of Tatius,' and another: 'I am a subject of Romulus'"—to exclude, in other words, the idea of different descent.

[2] The work of H. Schurtz, devoted to the "age-classes" and the secret men's unions during the barbarian stages of civilization (*Altersklassen und Männerverbände: eine Darstellung der Grundformen der Gesellschaft*, Berlin, 1902), which reaches me while I am reading the proofs of these pages, contains numbers of facts in support of the above hypothesis concerning the origin of guilds. The art of building a large communal house, so as not to offend the spirits of the fallen trees; the art of forging metals, so as to conciliate the hostile spirits; the secrets of hunting and of the ceremonies and mask-dances which render it successful; the art of teaching savage arts to boys; the secret ways of warding off the witchcraft of enemies and, consequently, the art of warfare; the making of boats, of nets for fishing, of traps for animals, and of snares for birds, and finally the women's arts of weaving and dyeing—all these were in olden times as many "artifices" and "crafts," which required secrecy for being effective. Consequently, they were transmitted from the earliest times, in secret societies, or "mysteries," to those only who had undergone a painful initiation. H. Schurtz shows now that savage life is honeycombed with secret societies and "clubs" (of warriors, of hunters), which have as ancient an origin as the marriage "classes" in the clans, and contain already all the elements of the

With regard to the above-mentioned work by E. Martin-Saint-Léon, let me add that it contains very valuable information concerning the organization of the trades in Paris—as it appears from the *Livre des métiers* of Boileau—and a good summary of information relative to the Communes of different parts of France, with all bibliographical indications. It must, however, be remembered that Paris was a "Royal city" (like Moscow, or Westminster), and that consequently the free mediæval-city institutions have never attained there the development which they have attained in free cities. Far from representing "the picture of a typical corporation," the corporations of Paris, "born and developed under the direct tutorship of royalty," for this very same cause (which the author considers a cause of superiority, while it was a cause of inferiority—he himself fully shows in different parts of his work how the interference of the imperial power in Rome, and of the royal power in France, destroyed and paralyzed the life of the craft-guilds) could never attain the wonderful growth and influence upon all the life of the city which they did attain in North-Eastern France, at Lyons, Montpellier, Nîmes, etc., or in the free cities of Italy, Flanders, Germany, and so on.

XI.—THE MARKET AND THE MEDIÆVAL CITY.

(To p. 190.)

In a work on the mediæval city (*Markt und Stadt in ihrem rechtlichen Verhältnis*, Leipzig, 1896),

---

future guild: secrecy, independence from the family and sometimes the clan, common worship of special gods, common meals, jurisdiction within the society and brotherhood. The forge and the boat-house are, in fact, usual dependencies of the men's clubs; and the "long houses" or "palavers" are built by special craftsmen who know how to conjure the spirits of the fallen trees.

Rietschel has developed the idea that the origin of the German mediæval communes must be sought in the *market*. The local market, placed under the protection of a bishop, a monastery or a prince, gathered round it a population of tradesmen and artisans, but no agricultural population. The sections into which the towns were usually divided, radiating from the market-place and peopled each with artisans of special trades, are a proof of that: they formed usually the Old Town, while the New Town used to be a rural village belonging to the prince or the king. The two were governed by different laws.

It is certainly true that the market has played an important part in the early development of all mediæval cities, contributing to increase the wealth of the citizens, and giving them ideas of independence; but, as has been remarked by Carl Hegel—the well-known author of a very good general work on German mediæval cities (*Die Entstehung des deutschen Städtewesens*, Leipzig, 1898), the town-law is not a market-law, and Hegel's conclusion is (in further support to the views taken in this book) that the mediæval city has had a double origin. There were in it "two populations placed by the side of each other: one rural, and the other purely urban;" the rural population, which formerly lived under the organization of the *Almende*, or village community, was incorporated in the city.

With regard to the Merchant Guilds, the work of Herman van den Linden (*Les Gildes marchandes dans les Pays-Bas au Moyen Age*, Gand, 1896, in *Recueil de travaux publiés par la Faculté de Philosophie et Lettres*) deserves a special mention. The author follows the gradual development of their political force and the authority which they gradually acquired upon the industrial population, especially on the drapers, and describes the league concluded by the artisans to oppose their growing power. The idea, which is developed in this book, concerning the appearance of the merchant guild at a later period which mostly corresponded to a period of decline of the city liberties, seems thus to find confirmation in H. van den Linden's researches.

XII.—MUTUAL-AID ARRANGEMENTS IN THE VILLAGES OF NETHERLANDS AT THE PRESENT DAY.

(To p. 250.)

The Report ot the Agricultural Commission of Netherlands contains many illustrations relative to this subject, and my friend, M. Cornelissen, was kind enough to pick out for me the corresponding passages from these bulky volumes (*Uitkomsten van het Onderzoek naar den Toestand van den Landbouw in Nederland*, 2 vols. 1890).

The habit of having one thrashing-machine, which makes the round of many farms, hiring it in turn, is very widely spread, as it is by this time in nearly every other country. But one finds here and there a commune which keeps one thrashing-machine for the community (vol. I. xviii. p. 31).

The farmers who have not the necessary numbers of horses for the plough borrow the horses from their neighbours. The habit of keeping one communal bull, or one communal stallion, is very common.

When the village has to raise the ground (in the low districts) in order to build a communal school, or for one of the peasants in order to build a new house, a *bede* is usually convoked. The same is done for those farmers who have to move. The *bede* is altogether a widely-spread custom, and no one, rich or poor, will fail to come with his horse and cart.

The renting in common, by several agricultural labourers, of a meadow, for keeping their cows, is found in several parts of the land; it is also frequent that the farmer, who has plough and horses, ploughs the land for his hired labourers (vol. I. xxii. p. 18, etc.).

As to the farmers' unions for buying seed, exporting vegetables to England and so on, they become universal. The same is seen in Belgium. In 1896, seven years after peasants' guilds had been started, first in the Flemish part of the country, and four years only after they were introduced in the Walloon portion of Belgium, there were already 207 such guilds, with a membership of 10,000 (*Annuaire de la Science Agronomique*, vol. I. (2), 1896, pp. 148 and 149).

INDEX

Richard Clay & Sons, Limited, London and Bungay.

Printed in Dunstable, United Kingdom

91219954R00211